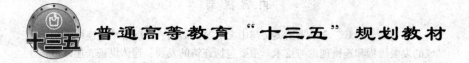

普通高等教育"十三五"规划教材

Continuous Casting Process and Technology
连铸工艺与技术

(双语版)

马胜强　杨军　编著

北　京

冶金工业出版社

2020

内 容 提 要

本书共分七章,全书采用英文与中文对照的形式,系统总结了连续铸钢领域的发展历程和连铸理论与技术,主要包括连铸的发展、连铸设备、连铸传热和凝固、连续铸钢工艺、铸坯缺陷和质量控制、连续铸钢新技术、特殊连铸工艺等内容。

本书可供高等院校冶金、金属加工等专业师生阅读,也可供相关专业从业人员参考。

图书在版编目(CIP)数据

连铸工艺与技术=Continuous Casting Process and Technology:双语版:汉文、英文/马胜强,杨军编著. —北京:冶金工业出版社,2020.4

普通高等教育"十三五"规划教材
ISBN 978-7-5024-8454-5

Ⅰ.①连… Ⅱ.①马… ②杨… Ⅲ.①连续铸钢—高等学校—教材—汉、英 Ⅳ.①TF777

中国版本图书馆 CIP 数据核字(2020)第 051912 号

出版人 陈玉千
地　　址　北京市东城区嵩祝院北巷 39 号　邮编　100009　电话　(010)64027926
网　　址　www.cnmip.com.cn　电子信箱　yjcbs@cnmip.com.cn
责任编辑　曾媛　美术编辑　郑小利　版式设计　孙跃红
责任校对　李娜　责任印制　李玉山
ISBN 978-7-5024-8454-5
冶金工业出版社出版发行;各地新华书店经销;三河市双峰印刷装订有限公司印刷
2020 年 4 月第 1 版,2020 年 4 月第 1 次印刷
787mm×1092mm　1/16;12.5 印张;298 千字;184 页
49.00 元

冶金工业出版社　投稿电话　(010)64027932　投稿信箱　tougao@cnmip.com.cn
冶金工业出版社营销中心　电话　(010)64044283　传真　(010)64027893
冶金工业出版社天猫旗舰店　yjgycbs.tmall.com

(本书如有印装质量问题,本社营销中心负责退换)

Preface

Iron and steel metallurgy is a pillar industry of national economy manufacturing industry, and the appearance of continuous steel casting technology has become a milestone of iron and steel metallurgy and material science and technology innovation. Although continuous casting technology has been invented for more than 100 years, its large-scale industrial application and rapid development is only in the past 40 years. It can be said that continuous casting technology is fast-changing in the development of contemporary iron and steel industry. In order to meet the needs of internationalization and development of continuous casting technology, a bilingual edition book of "Continuous Casting Process and Technology" (bilingual edition) was compiled with many efforts.

Compared with the advanced developed countries, the overall level of China's steel manufacturing industry needs to be further improved. Despite the remarkable technical progress in the field of continuous casting in recent years, efforts still need to be made in the aspects of the design of continuous casting machines and production lines, light downforce technology and metallurgical quality, the control of large cross section size billet and the process stability of thin slab. In the face of the fierce competition in the world steel market, we may develop our iron and steel industry only with the most state-of-art continuous casting technology as a support, constantly relying on the current intelligent control means, pioneering innovation, in order to stand out in the competition and meet the future steel industry's severe challenges.

So far, comprehensive and systematic bilingual books on continuous casting technology are rare. We have read a large number of original English works and teaching materials on continuous casting of iron and steel, and compiled this bilingual edition after careful selection. To making engineering students and researchers familiar with and master the technical vocabulary related to continuous casting as soon as possible, according to the basic principle and characteristics of continuous casting technology and molding, with the method of combination of English and Chinese, the book gives prominence to the continuous casting process and is introduced basic principle, process and related technologies, including continuous casting technology and equipment, with casting solidification heat transfer of continuous casting, continuous casting process, casting defects and quality control, etc. , and the new technologies of continuous casting processing are summarized.

This book can be used as a teaching and reading material for undergraduate and graduate students majoring in metallurgical engineering or material processing. It can also be used as a reference for related professional engineers and technicians.

The book is edited by Ma Shengqiang from the School of Materials Science and Engineering, Xi'an Jiaotong University, and Yang Jun from the School of Metallurgical Engineering, Xi'an University of Architecture and Technology. Ma is the editor-in-chief and also compiled the manuscript and proofread the book. Li Hui (Chapter 1), Wang Ruihu (Section 1, Section 2 in Chapter 6) and Zhang Dongliang (Chapter 7) from Xi'an University of Architecture and Technology were also helpful. In the process of publication, the book was funded by the National Natural Science Foundation of China (51771143), National Joint Engineering Research Center for Abrasion Control and Molding of Metal Materials (HKDNM201801) and Xi'an University of Architecture and Technology First-class Professional Sub-project Furnace Refining (YLZY0802K05).

This is the first attempt to compile bilingual books. It is not experienced enough, and there will be some omissions in the literature, or even errors and shortcomings in the book, and experts and scholars are invited to criticize and correct for it.

Authors

2020. 1

前　言

钢铁冶金是国民经济制造业的支柱性产业，而连续铸钢技术的出现已成为钢铁冶金和材料科学技术革新的一个里程碑。尽管连铸技术的发明已有100多年的历史，但其工业化大规模应用和飞速发展也只不过是近四十多年的事情。可以说，连铸技术在当代钢铁工业发展中更是日新月异。为了适应对连铸技术学习的国际化和发展需要，经过多方努力，我们合编了《Continuous Casting Process and Technology 连铸工艺与技术》（双语版）一书。

同先进发达国家相比，我国的钢铁制造业整体水平还有待进一步提高。尽管近些年连铸领域取得了显著的技术进步，但在连铸机及生产线设计、轻压下和冶金质量、大断面尺寸铸坯控制和薄板坯工艺稳定性等方面仍需努力。面对当前世界钢铁市场的激烈竞争，只有以最先进的连铸技术作为支撑，不断依赖当前智能化控制手段，开拓创新，才能在竞争中脱颖而出，迎接未来钢铁业的严峻挑战。

到目前为止，全面、系统的有关连铸技术的双语版书籍尚不多见。我们参阅了大量有关钢铁连铸的英文原版著作和教材，并经过仔细筛选后编写了这本双语版著作。为了使工科学生及研究人员尽快熟悉和掌握连铸相关的技术词汇，本书根据连铸工艺及成型的基本原理和特点，采用英文和中文相结合的方式，突出介绍了连铸过程相关的基本原理、工艺和技术，包括连铸工艺与设备、连铸传热与铸坯凝固、连续铸钢工艺、铸坯缺陷和质量控制等，并对连续铸钢新技术进行了总结。

本书可作为高等学校冶金工程和材料加工专业本科生及研究生教材，也可供相关专业工程技术人员阅读及参考。

本书由西安交通大学材料科学与工程学院马胜强和西安建筑科技大学冶金工程学院杨军编写，马胜强担任本书主编并对全书进行了统稿和校审工作。部分章节还得到西安建筑科技大学研究生李辉（第1章）、王瑞虎（第6章第1、2节）、张东亮（第7章）的帮助。本书的出版得到了国家自然科学基金

(51771143)、国家地方联合工程中心基金重点项目（HKDNM201801）和西安建筑科技大学一流专业子项目炉外精炼（YLZY0802K05）的资助，在编写过程中参考了国内外学者的有关成果，在此一并表示感谢。

首次尝试双语书籍编著，经验不足，错漏在所难免，敬请专家及读者批评指正。

作　者

2020 年 1 月

Contents

1 Introduction 1
  1.1 Description of Continuous Casting Process 1
  1.2 The Development and Evolution of the Continuous Casting 1
  1.3 The Superiority of the Continuous Casting 5

2 Machine Components of Continuous Casting 7
  2.1 Main Parameters of Continuous Caster 7
    2.1.1 Definition of a set of caster and strand 7
    2.1.2 Range of continuously cast sections 7
    2.1.3 Casting speed 8
    2.1.4 Arc radius 8
    2.1.5 Metallurgical length 9
  2.2 The Key Machine Components of Continuous Caster 10
    2.2.1 Ladle and ladle turret 10
    2.2.2 Tundish and tundish car 10
    2.2.3 Mould 13
    2.2.4 Dummy bar systems 17
    2.2.5 Strand support systems and secondary cooling 18
    2.2.6 Secondary cooling 21
    2.2.7 Strand straightening and strand withdrawal 26
    2.2.8 Product discharge and handling 29

3 Heat Transfer and Strand Solidification 31
  3.1 Mould Heat Transfer 31
  3.2 Secondary Cooling Zone Heat Transfer 37
    3.2.1 Solidification processing 37
    3.2.2 The heat transfer in secondary cooling zone 38
  3.3 Solidification of the Continuous Casting Strand 39
    3.3.1 The solidification process 39
    3.3.2 Solidification of the molten steel in mould 40
  3.4 Modeling of Continuous Casting 43
    3.4.1 Physical models 43

3.4.2 Computational models ... 45
3.4.3 Fluid flow models ... 46

4 Continuous Casting Process ... 49
4.1 Foreword ... 49
4.2 Treatment of Liquid Steel ... 49
4.2.1 The temperature control ... 49
4.2.2 The composition control ... 50
4.2.3 Impurities control ... 51
4.3 Tundish to Mould ... 52
4.4 Secondary Cooling ... 55
4.4.1 The control of secondary cooling ... 55
4.4.2 Strand containment ... 56
4.4.3 Strand withdrawal ... 56
4.4.4 Cooling water system ... 57
4.4.5 Casting speed control ... 58
4.5 Startup, Control of the Process and Problems ... 60
4.6 Mould Flux Powder Practice ... 60
4.6.1 Properties of mould powder ... 61
4.6.2 Effect of casting conditions on mould flux requirements ... 66
4.6.3 Mould flux requirements and design by steel grade ... 67

5 Defect and Quality Control of Continuous Casting Products ... 69
5.1 Foreword ... 69
5.2 Purity of Continuous Casting Products ... 70
5.2.1 The inclusions in the casting steel and its origin ... 70
5.2.2 Measures to decrease the inclusions in the casting steel ... 71
5.3 Surface Crack and Its Control ... 71
5.3.1 Slab surface defects and controlling ... 71
5.3.2 Surface defects in bloom and billet ... 73
5.4 Internal Defect and Its Controlling ... 75
5.4.1 Internal defects in slab ... 75
5.4.2 Internal defects in blooms and billets ... 76
5.4.3 The methods to decrease the internal defects ... 77
5.5 Shape Defects of Continuous Casting Products ... 79
5.5.1 Bulging deformation ... 79
5.5.2 Rhomboidity of casting billet ... 81
5.5.3 Deformation of round bloom ... 82
5.6 Breakout ... 82

		5.6.1	Cause	83

    5.6.1   Cause ·············· 83
    5.6.2   The methods to prevent the breakout ·············· 83
    5.6.3   Application of multivariate PCA for breakout prevention ·············· 83
  5.7 Effect of Spray Cooling on the Quality of Continuous Casting Semis ·············· 84

6 New Approaches in Continuous Casting of Steel ·············· 86
  6.1 Metallurgical Techniques of Tundish ·············· 86
    6.1.1   Heating technology of tundish ·············· 87
    6.1.2   Flow control in tundish ·············· 87
    6.1.3   Tundish flux ·············· 88
    6.1.4   Centrifugal flow tundish ·············· 89
  6.2 The Mould-level Control ·············· 90
  6.3 Hot Charging and Direct Rolling of Continuous Casting Slab ·············· 92
  6.4 Soft Reduction at the Final Stage of Solidification ·············· 93
    6.4.1   Effects of soft reduction ·············· 93
    6.4.2   Problem of attention in soft reduction ·············· 95
  6.5 Electromagnetic Stirring Techniques ·············· 96
    6.5.1   Historical development of electromagnetic stirring ·············· 96
    6.5.2   Effects of electromagnetic stirring ·············· 98
    6.5.3   Types of electromagnetic stirrers ·············· 99
    6.5.4   Metallurgical aspects of electromagnetic stirring ·············· 101

7 Special Continuous Casting Processes ·············· 103
  7.1 Horizontal Casting ·············· 103
  7.2 Beam Blank Casting ·············· 105
  7.3 Thin Slab Casting ·············· 107
    7.3.1   Problem ·············· 108
    7.3.2   Practice ·············· 109
  7.4 Strip Casting ·············· 109
    7.4.1   Twin-roll casting ·············· 111
    7.4.2   Thin strip casting of high-speed steels ·············· 117

目 录

1 引言 ······ 119
  1.1 连铸工艺简介 ······ 119
  1.2 连铸的发展及机型演变 ······ 119
  1.3 连铸的优势 ······ 121

2 连铸设备 ······ 122
  2.1 连铸机主要参数 ······ 122
    2.1.1 连铸机机数和流数的定义 ······ 122
    2.1.2 铸坯断面尺寸 ······ 122
    2.1.3 拉坯速度 ······ 122
    2.1.4 弧形半径 ······ 123
    2.1.5 冶金长度 ······ 123
  2.2 连铸机关键部件 ······ 124
    2.2.1 钢包回转台 ······ 124
    2.2.2 中间包和中间包车 ······ 124
    2.2.3 结晶器 ······ 126
    2.2.4 引锭杆系统 ······ 127
    2.2.5 铸坯支撑系统和二次冷却 ······ 128
    2.2.6 二次冷却 ······ 129
    2.2.7 铸坯矫直和拉坯装置 ······ 132
    2.2.8 铸坯输送和后续处理 ······ 134

3 连铸传热和凝固 ······ 135
  3.1 结晶器传热 ······ 135
  3.2 二次冷却区传热 ······ 137
    3.2.1 二冷区凝固 ······ 137
    3.2.2 二冷区传热 ······ 137
  3.3 铸坯的凝固 ······ 138
    3.3.1 凝固过程 ······ 138
    3.3.2 结晶器中钢液的凝固 ······ 138
  3.4 连铸模拟 ······ 140
    3.4.1 物理模拟 ······ 140
    3.4.2 计算模型 ······ 141

3.4.3 流体流动模型 …………………………………………………………… 142

4 连续铸钢工艺 …………………………………………………………………… 144
4.1 引言 …………………………………………………………………………… 144
4.2 钢液处理 ……………………………………………………………………… 144
4.2.1 温度控制 ………………………………………………………………… 144
4.2.2 成分控制 ………………………………………………………………… 145
4.2.3 夹杂物控制 ……………………………………………………………… 145
4.3 中间包到结晶器 ……………………………………………………………… 145
4.4 二次冷却 ……………………………………………………………………… 147
4.4.1 二冷控制 ………………………………………………………………… 147
4.4.2 铸坯夹持 ………………………………………………………………… 147
4.4.3 拉坯 ……………………………………………………………………… 148
4.4.4 冷却水系统 ……………………………………………………………… 148
4.4.5 拉速控制 ………………………………………………………………… 149
4.5 连铸的开浇、控制及问题 …………………………………………………… 149
4.6 结晶器保护渣 ………………………………………………………………… 150
4.6.1 结晶器保护渣的特性 …………………………………………………… 150
4.6.2 连铸工艺条件对结晶器保护渣的影响 ………………………………… 152
4.6.3 钢种对结晶器保护渣的影响 …………………………………………… 152

5 铸坯缺陷和质量控制 …………………………………………………………… 154
5.1 前言 …………………………………………………………………………… 154
5.2 铸坯洁净度 …………………………………………………………………… 155
5.2.1 铸坯夹杂物及其来源 …………………………………………………… 155
5.2.2 减少夹杂物的措施 ……………………………………………………… 155
5.3 表面裂纹及其控制 …………………………………………………………… 155
5.3.1 板坯表面缺陷及其控制 ………………………………………………… 155
5.3.2 大方坯和小方坯的表面缺陷 …………………………………………… 157
5.4 内部缺陷及控制 ……………………………………………………………… 158
5.4.1 板坯内部缺陷 …………………………………………………………… 158
5.4.2 大方坯和小方坯的内部缺陷 …………………………………………… 158
5.4.3 减少内部缺陷的途径 …………………………………………………… 159
5.5 铸坯形状缺陷 ………………………………………………………………… 160
5.5.1 鼓肚变形 ………………………………………………………………… 160
5.5.2 小方坯脱方 ……………………………………………………………… 161
5.5.3 圆坯的变形 ……………………………………………………………… 162
5.6 漏钢 …………………………………………………………………………… 162
5.6.1 漏钢的原因 ……………………………………………………………… 162

5.6.2　防止漏钢的措施 …………………………………………………… 163
　　5.6.3　多变量PCA防止漏钢 ……………………………………………… 163
　5.7　喷水冷却对铸坯质量的影响 ………………………………………………… 163

6　连续铸钢新技术 ………………………………………………………………… 165
6.1　中间包冶金 …………………………………………………………………… 165
　　6.1.1　中间包加热技术 ……………………………………………………… 165
　　6.1.2　中间包流动控制 ……………………………………………………… 166
　　6.1.3　中间包保护渣 ………………………………………………………… 166
　　6.1.4　离心流中间包 ………………………………………………………… 167
6.2　结晶器液面控制 ……………………………………………………………… 167
6.3　板坯热装和直接轧制 ………………………………………………………… 168
6.4　凝固末期轻压下 ……………………………………………………………… 168
　　6.4.1　轻压下的效果 ………………………………………………………… 169
　　6.4.2　轻压下需要注意的问题 ……………………………………………… 169
6.5　电磁搅拌技术 ………………………………………………………………… 170
　　6.5.1　电磁搅拌的发展 ……………………………………………………… 170
　　6.5.2　电磁搅拌的作用 ……………………………………………………… 171
　　6.5.3　电磁搅拌器类型 ……………………………………………………… 172
　　6.5.4　电磁搅拌的冶金效果 ………………………………………………… 173

7　特殊连铸工艺 …………………………………………………………………… 174
7.1　水平连铸 ……………………………………………………………………… 174
7.2　异型坯连铸 …………………………………………………………………… 175
7.3　薄板坯连铸 …………………………………………………………………… 176
　　7.3.1　存在的问题 …………………………………………………………… 176
　　7.3.2　实际生产 ……………………………………………………………… 177
7.4　带钢浇注工艺 ………………………………………………………………… 177
　　7.4.1　双辊式浇注 …………………………………………………………… 178
　　7.4.2　高速钢薄板带连铸 …………………………………………………… 182

参考文献 ……………………………………………………………………………… 183

1 ◆ Introduction

1.1 Description of Continuous Casting Process

The basic principle of the continuous casting process for steel is based on teeming liquid steel vertically into a water cooled copper mould which is open at the bottom. A brief description of a modern continuous casting machine at this stage will help the reader to appreciate the various aspects of the process. Fig. 1-1 shows a general layout of a modern continuous slab casting machine, showing the ladles in the ladle turret. This turret revolves so that a full ladle of steel can be brought to the casting position quickly to enable continuity of casting.

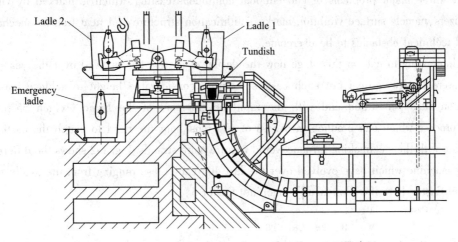

Fig. 1-1 General layout of a modern continuous casting plant

The liquid steel is initially teemed from the steelmaking vessel into the ladle and following any appropriate secondary steelmaking processing the ladle is lifted by crane onto the continuous casting machine and supported by either a ladle car or ladle turret. The liquid steel is then poured from the ladle into a tundish by way of a sliding gate valve mechanism and the stream is protected by a refractory tube to avoid any reoxidation from the atmosphere. The liquid stream between tundish and mould is again protected from the atmosphere by a refractory tube. The strand becomes completely solid after passing several meters down the machine and then straightened by the use of rollers at the position where it becomes horizontal and is withdrawn from the machine by power-driven pinch rolls.

1.2 The Development and Evolution of the Continuous Casting

Continuous casting(CC) of steel, as an industrialized method of solidification processing, has a relatively short history not much longer than oxygen steelmaking. During the rather lengthy incubation in the precursory periods, i. e., before the 1950s, important development stimuli came from the nonferrous industry, which had applied CC processes already—in particular, by the traveling mould principle—using casting wheels and/or belts to overcome mould friction. Later, genuine ideas emanating from steelmakers added various milestones to the driving of CC application to steel, though primarily by a process based on a stationary, oscillating mould. In fact, the CC ratio for the world steel industry, approaching 96.2% of crude steel output in 2018, attained a mere 4% in 1970.

Continuous casting has been developed very rapidly. Dr. Wolff summarized the development of CC as six stages of continuous challenge, innovation and remarkable achievements which include the experimental exploration in the 1940s, the industrialization in the 1950s, the revolution in the 1960s, the two energy crisis in the 1970s, more mature technology in the 1980s, and high productivity, high-speed casting, high-quality products, energy saving and environmental protection, improved working conditions, process and quality control technology, and near net shape casting (thin slab, thin strip) since the 1990s. Although today's continuous casting has reached a fairly mature level, the three major problems of conventional continuous casting, which is marked by vibrating crystallizers, namely surface vibration marks, solidification structure and heat transfer mechanism, are still technical obstacles to be overcome.

It is interesting to note at this stage how the design of casters has evolved over the years. Early casters were totally vertical but such casters required considerable height to achieve reasonable production rates per strand and with the rapid development of the Basic Oxygen Steelmaking (BOS) process which can produce in excess of 400 tons/hour the need to match the casting machine rate with the steelmaking furnace would require more strands. Fig. 1-2 shows the different designs of machine which have evolved over the last 40 years, these ranging from the totally vertical machine (Caster 1) to the "low head" machine (Caster 5).

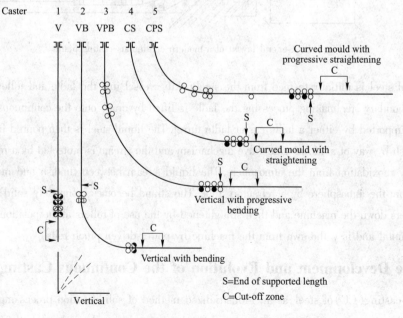

Fig. 1-2 Principal types of continuous casting machines

1.2 The Development and Evolution of the Continuous Casting

In 1965, the continuous casting machines were very simple. 80% of the casters, for slabs, blooms and billets, were vertical machines. Curved machines then took over and in 1975 80% of the slab casters and 70% of the bloom and billet casters were of the curved type. This trend continued to progress but towards more complex geometry, with the application of progressive bending and straightening which in 1984 was used in 30% of the slab casters and in 20% of the bloom and billet casters.

(1) Vertical caster.

Casting vertically has certain inherent technical advantages over the now much more popular horizontal casting processes. The symmetry of cooling ensures a uniform and predictable crystal growth pattern and uniform axial loading on the freshly solidified shell as it is pulse withdrawn from the die. On tube or hollow section casting the process has particular merit as it eliminates the end-of-run spear and therefore an acceptable product is produced to the end of the cast. This latter advantage is important when casting precious metals.

(2) Vertical with bending caster.

A radius configuration of a strand guide of a vertical bending caster comprising:

1) A bending zone adjoining the vertical section.

2) A transition straightening zone adjoining the bending zone.

3) A main radius guide zone adjoining the transition straightening zone.

4) A transition final straightening zone arranged between the main radius guide zone and the horizontal section.

Wherein the bending zone has at least one first region with a comparatively stronger bend and defined by a curvature formed by a series of continuously smaller, starting from an infinite radius of the vertical section, radii, and a second region adjoining the first region and having a comparatively smoother bend in comparison with that of the first region.

Wherein the transition straightening zone, which forms a transition between the bending zone and the main radius guide zone, is defined by a curvature with continuously increasing lengths of curvature-forming radii, and wherein the first and second regions of the bending zone and the transition straightening zone have each a clothoid shape.

In a vertical-bending caster, the straight mould and the straight segments of secondary cooling section can provide and good symmetrical cooling environment for solidification of liquid steel and strand with liquid core, and also is a fundamental condition to ensure the mould and secondary cooling section with a same level uniform heat flux. It is also an important basis to cast high quality products. The gravity from the vertical-bending caster can replace the withdrawing force from the withdraw machine as possible and then change the forced condition and reduce the outer force (withdrawing force) to slab. Therefore, the slab quality can be improved (see Fig. 1-3).

(3) Curve type caster.

The vertical caster is the natural machine design, casting with gravity and also assuring a symmetric macrostructure; but caster productivity is severely limited by machine height. Hence, several efforts in CC history are noteworthy to extend machine length at low building height by strand ben-

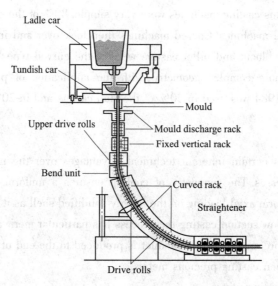

Fig. 1-3 Vertical with bending caster

ding and straightening, e. g. , the billet caster by Rowley and a more advanced proposal by Tarquinee and Scovill, which even includes strand in-line sizing after temperature equalization, a design concept that subsequently had been realized first by U. S. Steel for the South Works pilot and the Gary Works No. 1 slab caster, respectively (see Fig. 1-4). Commonly, the casting machine radius of the curved caster is 9-12m. The caster which developed by CERI can product bloom with a diameter of 1000mm. That casting machine has a radius of 17m and a length of 37.6m.

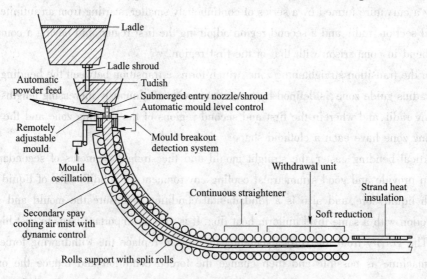

Fig. 1-4 Curve type continuous caster

In a curved casting machine, the strand exits the mould vertically (or on a near vertical curved path) and as it travels through the spray-chamber, the rollers gradually curve the strand towards the horizontal.

Efforts to reduce bulking-height first led to continuous-casting system in which molten metal passed into a vertical mould and solidified completely before being bent or where the strand has been in the liquid phase and later to a curved mould and is the system most used today. Vertical system and those in which the strand is bent when completely solidified have long straight liquid phases and can lead to unacceptably high capital outlay.

However, these system have metallurgical advantage from the point of view of maintenance. A vertical system in which the strand is bent while still in the liquid phase has the advantage that the building need not be as tall as when the strand is bent after solidification; however, the liquid-phase bending system represents a compromise between the costs of capital outlay and of maintenance.

To prevent inner cracking, several rules for caster design, based on critical strain and strain rate at the solid/liquid (s/l) interface, had been developed, which has led to distinct bending and straightening zones extending over several roller pairs.

(4) Horizontal caster.

The majority of continuous casting installations in use today operate in the horizontal mode. In a true "Horizontal Casting Machine" (image is not available) the mould axis is horizontal and the flow of steel is horizontal from liquid to thin shell to solid (no bending). In this type of machine, either strand oscillation or mould oscillation is used to prevent sticking in the mould. The reason for this is mainly logistic, based on ease of product handling and to some extent safety in operation. There are, of course, inherent problems applying horizontal as opposed to vertical casting mainly related to gravity-induced directional cooling; however, in most cases these difficulties can be accommodated (see Fig. 1-5).

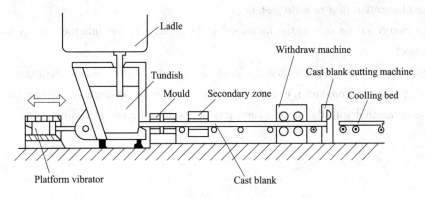

Fig. 1-5 Horizontal continuous caster

1.3 The Superiority of the Continuous Casting

The continuous casting of steel gives considerable advantages when compared with ingot casting. Fig. 1-6 shows the two process routes.

Ingot casting involves more processes with at least one extra heating and rolling process to produce similar semi products which are produced directly from the continuous casting process these

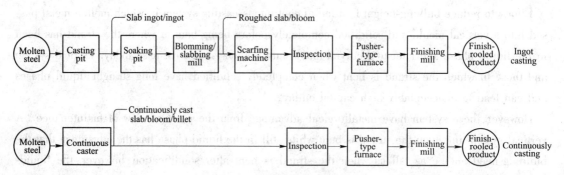

Fig. 1-6 Ingot and continuous casting process routes

being either billets, blooms or slabs.

The main advantages of the continuous casting process over the ingot casting route are listed as follows:
- improved yield;
- reduced energy consumption;
- savings in manpower;
- improved product quality;
- lower emissions harmful to the environment and plant operators;
- reduced stock levels and shorter delivery times;
- reduction in capital costs for new steel plants.

Reduction in energy consumption when comparing the continuously cast route to the ingot route arises due to the following:

(1) The elimination of a reheating stage;

(2) The energy saving due to the increased yield because of the inherent energy contained in the liquid steel.

The ingot process route requires that the stock needs to be heated both in the ingot form and after the rolling to a semi product i. e. slab, bloom or billet. In the continuous casting route, the first reheating is eliminated with the as-cast semi product being reheated for rolling.

2 Machine Components of Continuous Casting

2.1 Main Parameters of Continuous Caster

2.1.1 Definition of a set of caster and strand

Continuous casting machine is the basic and most important factor for the well working of continuous casting process. A set of continuous cast machine is definited as that have independence drive system and independence work system, and can keep working when other machines appear malfunction. One continuous cast machine can build up a machine or machines. One continuous cast machine can cast one or more casting semis at the same time, this semi is named as strand. A continuous cast machine which has one set of withdrawing device and driving set itself is named a caster. A set of continuous casting machine may have one or more than one withdrawing device and driving set and forms one or more than one strands, so we get one caster with one strand, one caster two strand, two caster with two strand.

2.1.2 Range of continuously cast sections

Concerning the types of continuous casting products, Table 2-1 indicates the different and common kind of products employed in the steel production process. According to the shape and size of continuous casting semis, the products can be classified as billet, bloom, round, slab and beam blank depending. In some cases, overlaps occur where the moulds on a particular machine can be changed to cast other shapes; for example, billets or blooms, blooms or small slabs, and blooms or rounds. In addition, machines exist where special shapes, such as rectangles and dog-bone structural sections can be cast as well as billets or blooms.

Table 2-1 Cast Cross Section of Typical Continuous Casters

Type	Max. cross section(mm)	Min. cross section(mm)	Regularity cast cross section(mm)
Slab	150×3250 700×2400	120×600	150×300 ~ 1000×1800
Bloom	600×600	200×200	250×250 ~ 450×450 240×280 ~ 400×560
Billet	180×180	55×55	120×120 ~ 150×150
Round	φ1000	φ45	φ200 ~ φ300

Low alloy steels containing 0.18 to 0.28 percent carbons are said to be difficult to cast unless secondary cooling is strictly controlled. However, substantial tonnages of grades such as SAE 8620 are cast by many works. Several plants are attempting to cast boron steels continuously.

There are also potentially important applications for stainless steel plates(austenitic), stainless steel sheet(austenitic, ferritic, martensitic), silicon steel for electric sheet, quenched and tempered steels, case-hardening steels, spring steels and weathering steels.

2.1.3 Casting speed

Casting speed is named that one strand of a continuous caster cast the length of blank in unit time. Obviously, increase the casting speed will improve the product capacity of continuous caster. So, casting speed is a very important parameter for continuous casting machine, which will affect the solidify speed of molten steel and internal quality of cast blank.

Assuming that the solidification coefficient $K=27$ for slab caster or 30 for bloom/billet caster and that the actual lengths employed (the metallurgical length, L) are 15, 20, 25 etc. meters, then the casting speed "v" is related to the slab thickness "d" by:

$$v = \frac{K^2 L}{(d/2)^2} \tag{2-1}$$

It is found that this formula has, in practice, reasonable predictive power; actual and calculated speeds follow a similar relationship and are almost equal. Steel grades can affect the speed due to change in "K", e.g. stainless steels are cast in smaller slab thickness and casting speeds generally lie between 0.8 and 1.0m/min.

For bloom/billet casers the deviations are much smaller than for slab caster, the range from maximum to minimum casting speed being a factor. Notable exceptions relate to US Steel's plants. Billets for seamless tubes are cast at a slow speed (1.5m/min) because quality requirements are high.

Looking at productivity in tons/minute/strand as a function of cross section of strand, the tonnage through slab casters naturally is much greater because of the larger cross-sectional area for a similar thickness. These data have been calculated in terms of the minimum, average and maximum speeds derived from the statistics provided.

2.1.4 Arc radius

Arc radii is the positive camber radii of casting blank, unit is m. It is an important parameter, which confirms the highness of a continuous cast machine and the thickness range of casting blank when casting.

Basic arc radii can be confirmed by experience formulary, and is the minimum arc radii of continuous caster.

$$R \geqslant cD \tag{2-2}$$

Where R is the arc radius of continuous caster; D is the thickness of casting blank; c is coefficient. As a rule, for billet caster, R value is 30-40m, bloom is 30-50m and slab is 40-50m.

Arc radius can be affirmed by theoretical calculation as well.

We request the cast blank must solidify completely before enter withdrawal straightening stands:

towards the continuous caster, which do not adopt multi-pairs withdrawal straightening rolls. The length of cooling zone can be calculated by follow:

$$L_c = \frac{2\pi R\alpha}{360} + h \tag{2-3}$$

In this formulary, R is the arc radii of cast blank center; α is the included angle of the arc radii center horizontal line and the axis of the first pairs rolls; h stands for the distance between the arc radii center horizontal and the molten steel fluid level in the mould. When we adopt the curved mould:

$$h = \frac{L}{2} - 0.1 \tag{2-4}$$

Here, L is the length of moulds. When we adopt the vertical mould, h is the length of straightway in second cooling zone.

The arc radius of continuous caster must assure the length of cooling zone must equal to or longer the length of liquid center. Therefore, the arc radius of continuous caster can be decided as follows:

$$R \geq \left(\frac{D^2}{4K^2}v - h\right) \times \frac{57.3}{\alpha} \tag{2-5}$$

Where D is the thickness of cast blank, mm; v is casting speed, m/min; K is synthesis solidification coefficient, mm/min$^{\frac{1}{2}}$; α is the included angle of the arc radii center horizontal line and the axis of the first pairs rolls.

2.1.5 Metallurgical length

The practicality length, from fluid level of molten steel in crystallizer to the axis of the last roller of withdrawal straightening strands, is named metallurgical length. It symbols the maximum limit position of casting blank liquid core depth. So, casters' metallurgical lengths have close relation with casting blank liquid core. The length liquid core of casting blank is from fluid level of molten steel in mould to the position where molten steel have been solidified. Liquid core length can be calculated by solidification law, the thickness of casting blank complete solidifies is S:

$$S = \frac{D}{2} = K\sqrt{L} = K\sqrt{\frac{L_1}{v}} \tag{2-6}$$

After arrangement, we can gain liquid core length L_1

$$L_1 = \frac{D^2 v}{4K^2} \tag{2-7}$$

Where L_1 is liquid core length, m; D is the thickness of casting blank, mm; v is casting speed, m/min; K is synthesis solidified coefficient, mm/min$^{\frac{1}{2}}$.

When design casting machine, we will think over not only the casting blank achieve the casting speed, but also the development of continuous casting technology after continuous cast machine be used, and the possible advanced casting speed further more. So, the continuous casters' metallurgical length can be gained by the maximum thickness of casting blank and maximum casting speed.

Essentially, metallurgical length is liquid core length. But because the practicality casting speed always lower than maximum designed casting speed, so liquid core length always lower than the metallurgical length of continuous casting. In addition, casting speed often change when casting, and liquid core length also change. Casting speed faster, the liquid core will longer, and the continuous caster is unalterable, is definite value.

2.2 The Key Machine Components of Continuous Caster

2.2.1 Ladle and ladle turret

Ladle turret: To ensure continuous operation, an extremely quick ladle change is needed, and the ladle turret is the preferred equipment for this duty, giving the fastest ladle change. A ladle is exchanged in around one minute, and total steel flow interruption to the tundish is around three minutes. This is the most popular form of ladle handling for sophisticated high-productivity machines. In their simplest form, ladle turrets are designed for fixed-height 180 degrees rotation of the ladle from receiving position to casting position.

Ladle turret variants: There are a large number of variants of ladle turret designs. Fixed arm or lift/lower arm turrets are applied as well as independent or common arm slew. The most popular high-productivity turret is the so-called "butterfly" type.

The butterfly turret handles a ladle on each pair of arms, which always rotate opposite to each other but are able to independently lift lower to provide ideal clearances of the ladle above the tundish at casting position. Such turrets are available for all ladle sizes. A further variant for special situations where the ladle has to approach at 90 degrees to casting direction is the "C" arm turret.

2.2.2 Tundish and tundish car

2.2.2.1 Tundish

The tundish as a reservoir holds enough liquid metal to provide a continuous flow to the mould, even during an exchange of ladles, which are supplied periodically from the steelmaking process. The inlet from the ladle is usually through a ceramic shroud, and the outlet(s) are arranged to match the mould positions below.

The tundish can also serve as a refining vessel to float out detrimental inclusions into the slag layer. The surface of the liquid steel in the tundish is protected from oxidation and heat loss by the application of a powdered synthetic slag, which then melts to form a barrier layer at the liquid steel surface. If solid inclusion particles are allowed to remain in the product, then surface defects such as "slivers" may form during subsequent rolling operations, or they may cause local internal stress concentration, which lowers the fatigue life.

The tundish serves the following purposes:

(1) To act as a reservoir sufficient to allow the exchange of ladles to enable sequence casting.

(2) To act as a flotation vessel to facilitate inclusion removal from the steel before entering the mould.

(3) To distribute the steel to multiple strands.

In order to fulfil these requirements, certain design principles are normally followed (see Fig. 2-1):

(1) A liquid steel average residence time in the order of 5-10 minutes to promote inclusion floatation and cleanness without incurring excessive temperature loss.

(2) A minimum depth of liquid steel of 2 feet(600mm) above the outlet(s) to avoid vortexing, which could entrain tundish cover slag.

(3) An operational capacity that allows for a normal ladle change before the tundish drains down to the minimum level, while casting is maintained at the normal casting speed.

(4) The elimination of "dead" volumes of low or zero flow.

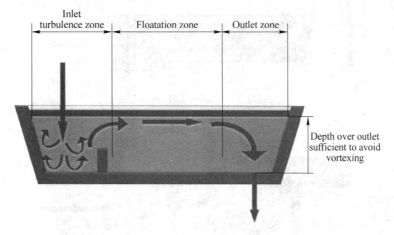

Fig. 2-1 Tundish design principles

There is usually a section of tundish at the inlet position that is designed to suppress turbulence, followed by a section that is designed to promote the rising of the liquid steel flow. The surface is covered by a liquid slag layer, which absorbs inclusions brought to the surface by the rising liquid steel flow and insulates the liquid steel, preventing excessive heat loss. The surface of the tundish may be additionally sealed with argon, either for the initial fill period before the surface powder is added or, alternatively, throughout casting operation if the tundish lid is completely sealed. The tundish shape, size and configuration of internal flow control devices such as dams, weirs and walls can be optimized using either physical water modeling or Computation Fluid Dynamics(CFD) modeling.

Fig. 2-2 shows one example in which CFD has been used to model the temperatures throughout the tundish for a range of flow control configurations. Similar modeling can also be used to estimate steel temperature variation—for example, in billet casting machine tundishes, which can distribute the steel to six or more individual strands, or for the evaluation of average residence times, dead volumes and inclusion flotation. Where long casting times are used, it is difficult to maintain the desired superheat in the tundish over the duration of the cast. To alleviate this problem, tundish heating can be applied (see Fig. 2-3). Such systems are applied in special cases and are generally based on plasma heating. One-and two-torch variants of plasma systems are applied. Plasma heating requires its own support systems, including power supply, water-cooling and manipulator for the torch(es).

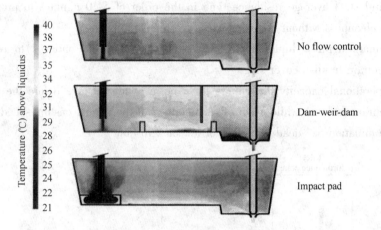

Fig. 2-2　Tundish CFD modeling

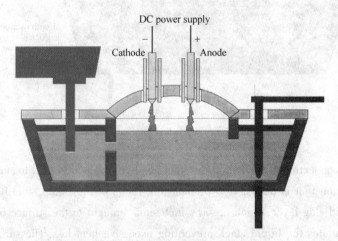

Fig. 2-3　Tundish plasma heating

For combination casting, the specific design arrangements start at the tundish. Multiple outlets are provided in the tundish shell, and the actual outlets utilized for a particular cast are selected according to the configuration being cast. Those outlets not in use are simply blocked with the refractory lining of the tundish.

In the simplest case, the tundish is located on a fixed stand at the casting position. Internal refractory life is typically limited to 10-20 hours casting. If there is a desire on modern casters to cast for longer than this, it is necessary to change tundishes during a cast; this is called a flying tundish change. Rapid tundish changing is required for this operation in order to minimize the time during which the strand is stopped. Longitudinal travel tundish cars are most commonly used to achieve this.

The main advantage of tundish cars is that on slab casters with two tundish cars, they have the facility to cast even when one of the cars is off line due to damage or maintenance. Tundish cars usually incorporate long and cross-travel motions as well as lift/lower. These motions may be hydraulically or electromechanically actuated. The alignment of the tundish refractory outlet tube to the mould wall can be critical.

When the tundish arrives at the mould, the tundish position may require a small adjustment in order to centralize the tube between the mould wide sides. On tundish cars, this cross-travel is commonly done by a simple manual hydraulic pumping system. For longer sequence casts and to reduce the number of flying tundish changes, there is a growing demand to change the tube during casting. Tube change mechanisms can therefore be incorporated, which allows the tube to be changed in a matter of seconds with minimal interruption to the casting operation.

As well as reducing the number of tundishes required during a cast, the tube change device also increases caster yield by reducing the number of tundish skulls and provides an emergency shutoff facility.

2.2.2.2 Tundish car

In addition to the gantry type tundish car, there are semi-gantry and cantilever type alternatives. These generally provide better access for the mould operator but do require a structure above the casting platform to support the high-level wheels. The cantilever car removes rails from the casting floor level, which is beneficial if open pouring is practiced. This structure gives less clear access for load movements by overhead crane, e. g. , for equipment changing, and so does require more handling care. The cantilever car is usually used in conjunction with the lighter tundishes employed on billet casters.

Flying tundish exchanges can generally be completed in around three minutes, with an actual strand(casting speed) stoppage of about 90 seconds. An alternative to the tundish car is the tundish turret, similar in principle to that used for ladle handling. This arrangement is compact but can lead to quite a congested area close to the casting mould. For this reason, the tundish turret is most often used for single strand slab casters where the size of the equipment does not need to be very large, or the casting floor space is limited.

2.2.3 Mould

The mould is the only mechanical part of a caster that is exposed to molten steel. It is probably the most important part of the machine and has to operate under severe conditions. It needs to create a homogeneous shell by efficient uniform heat transfer. The mould also needs to be long lasting, be capable of rapid change of section sizes, and require the minimum of maintenance effort.

Continuous casting moulds are all cooled by high quality water, often demineralized, supplied from a recirculating system. The design and fail-safe systems are usually arranged to provide a minimum water flow velocity in the cooling channels of 8m/s. Moulds are invariably tapered internally to accommodate contraction of the steel but the amount of taper depends on the section sizes and casting speeds involved.

Fig. 2-4 shows the basic construction of a billet(a), bloom(b) and slab mould(c) respectively. The copper moulds are contained by steel backing plates with water inlet and outlet manifolds at the bottom and top of the mould respectively.

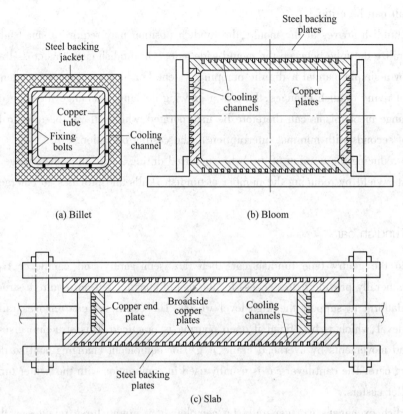

Fig. 2-4 Mould constructions for billet, bloom and slab casters

The water-cooling grooves are machined in the back of the copper plates from top to bottom in slab and bloom moulds the dimensions of these being about 15mm deep and 5mm wide. In billet moulds the water channel is usually a parallel gap between the tubular copper mould and the backing plate.

To ensure a thin boundary layer at the copper surface and hence no nucleate boiling, a high Reynolds number is required in these water cooling grooves which results in a need for the water velocities being greater than 8m/s.

The following are the two main mould types. These are:

(1) Tubular Moulds. These are frequently used for casting small sections such as billets. The copper tube is surrounded by the water-cooling jacket and, although easily deformed, the tube can be quickly exchanged or straightened. The maximum practical size is about 230mm square, or 430mm diameter for rounds castings but they are normally less than 200mm across. The larger sizes have greater wall thicknesses of about 20mm and on small sizes.

(2) Plate Moulds. These are assembled from four copper plates of 40 to 60mm thick. The cold faces are grooved and covered with a steel backing plate. The cooling water passes through these

grooves or, in an alternative design, through circular cooling channels machined in the copper. These moulds usually enable the narrow faces to be adjusted for different widths and these mechanisms can in some cases now be operated during casting.

The copper plates in bloom and slab moulds are usually between 50 and 60mm thick when new and about 40mm thick at the end of their lives. Usually several machinings of the face are carried out during the plate life.

2.2.3.1 Mould length

The normal mould length was, until recently, 700mm, but the range extends from 500 to 1,200mm. The most recent trend has been towards 900mm moulds to provide an increased solidified thickness at the mould outlet when casting at higher speeds.

2.2.3.2 Mould materials

The mould material must rapidly transmit the heat from the solidified steel to the cooling water and hence good thermal conductivity is essential. Copper and copper alloys are invariably used but it is necessary to minimize distortion from thermal stress. Silver, chromium and zirconium alloying additions are used because of their improved high temperature properties; Fig. 2-5 give details. In some cases, the working face of the mould is plated to minimize wear. This is claimed to reduce star cracks formed when copper adheres to the solidified shell but many plants, particularly in Europe, operate successfully without plating.

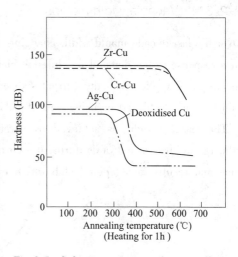

Fig. 2-5 Softening resistance of copper alloys

Various methods of plating the copper with nickel and chromium have been developed. One technique uses a thick layer so the mould can be reused after surface dressing. Other techniques taper the coating or use a two-stage plating method, the intention being to minimize wear at the lower part of the mould. Another technique uses nickel iron plating and the increased hardness doubles the wear resistance. Mould plating is most common in Japan and finds only limited application elsewhere.

2.2.3.3 Mould oscillation

The original idea of a reciprocating motion to prevent sticking between the shell and the mould is attributed to Junghans. With a few exceptions, the mould oscillation cycle is sinusoidal but, in every case, the downward velocity exceeds the casting speed for part of the cycle. During this time, (termed the negative strip time or heal time), sticking between the mould and the shell is overcome.

Mould oscillation is essential for the elimination of breakouts and under carefully controlled conditions the breakout rate can be virtually zero. The movement for mould oscillation is derived from a motor driven cam but hydraulic devices have been developed. The design of the structure, bearings and lever arms is critical since the stroke length must remain equal at different points on the mould and only very small horizontal or radial movements of less than 0.2mm can be tolerated.

For best results the mounting points of the oscillation system should be separated from the casting floor and machine frame. Defective oscillation will result in increased breakout rate and surface defects on the strand. Recent work has shown that there can be significant improvements to surface quality by operating with small heal times. This is usually achieved with small stroke lengths, down to 4mm on slab machines and down to 8mm on billet machines and oscillation frequencies of 200 cycles/minute(cpm) or greater compared to the more usual 100 or 120 cpm. These higher frequencies and small stroke lengths, have shown benefits on some stainless-steel casters and are becoming more common elsewhere, and place a greater demand on the design and upon the engineering standards for trouble free operation.

2.2.3.4 Variable width moulds

Over the last decade mould width changing during casting on slab machines has been established in a response to the demand for different slab widths without interruption of a sequence cast. The technique is applied in many current slab casters. A maximum width changing speed of 200mm/min has been achieved by using a carefully chosen sequence of moving the narrow plates.

The variable width is achieved by careful movement of the narrow faces which are power adjusted inwards or outwards during the casting process. The adjustment is made over a period of time and results in a tapered slab which may need special attention during reheating. Fig. 2-6

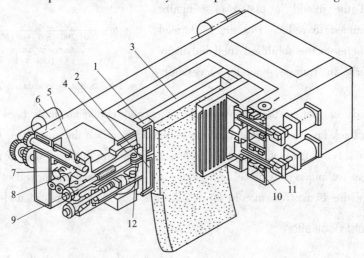

Fig. 2-6 Width adjustable mould with horizonally split narrow faces

Mould construction: 1—Top narrow faces; 2—Bottom narrow faces; 3—Broad faces

Taper adjustment system: 4—Cage with rotary segment; 5—Cam; 6—Drive

Width adjustment and measuring system: 7—Position indicator (pulse generator); 8—Positioning motor; 9—Spindle

Mould clamping system: 10—Release mechanism for width change; 11—Mould clamping device for casting;

12—Marrow face locking mechanism

shows the main components required for such adjustments.

It is critical during the width change that the taper of the end plate is accurately controlled, the taper varying as the width is changed.

It is necessary to have inclinometers installed to measure the taper continuously and Fig. 2-7 shows the sequence of events for both changing to a wider or narrower slab width.

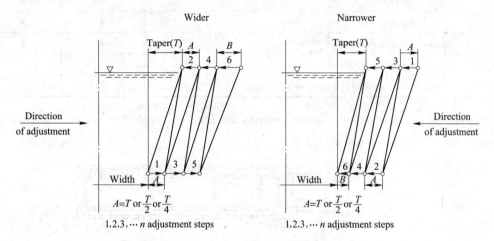

Fig. 2-7 Stages in a width adjustment operation

It is reported that width adjustment during casting can result in an increase in production of 30%-50%, a reduction in refractory costs by 30%-50%, an increase in yield of 0.3%-0.5% and significant savings in energy. The energy savings are realized by the fact that width changing increases the ability to hot charge and/or direct roll since it is then possible to match the rolling schedule.

2.2.4 Dummy bar systems

To allow cast start, a "dummy bar" must be provided onto which to cast the strand at the start of the cast. The dummy bar chain is driven up by the "withdrawal" rolls and the head is placed in position which extends slightly into the bottom of the mould. The dummy bar head is shaped in a claw like fashion so that when liquid steel enters the mould it solidified around the "claw" and when the mould is filled withdrawal is started and the dummy bar commences to withdraw the partly solidified steel from the mould. When the dummy bar head and the leading end of the strand exit the machine, the head is disconnected and the dummy bar chain withdrawn separately and parked in ambush. For a simple vertical or small section (billet) curved caster, the dummy bar can be exactly what the words say—a solid, rigid bar similar to the strand cross-section. For larger sizes and for more complex machine profiles, a flexible chain dummy bar is most frequently used. Less common but also used for thinner flat sections is a plate-or belt-type bar. In this case, a thin plate runs through the whole length of the bar, and localized packing pieces are added to give the bar the required thickness but still allow it to flex as it travels through the machine. A comparison of applications for the various dummy bar types is given in Table 2-2.

Table 2-2 Comparison of dummy bar types

Item	Rigid type	Chain type	Belt type
Minimized turnaround time	●		
Section<7in(180mm)	●		
Section>7in(180mm)		●	●
Bay height restriction		●	●
Strand center distance<37.5in(950mm)		●	●
Strand center distance>37.5in(950mm)	●		
Reduced No. of rolls in casting bow	●		

For combination caster, multiple leading lengths are fitted to match the number of combination strands being cast, as illustrated in Fig. 2-8.

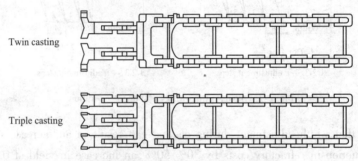

Fig. 2-8 Combination casting dummy bars

The dummy bar head is usually arranged to provide automatic disconnection of the dummy bar from the strand once the bar is clear of the strand support and withdrawal devices. In practice, the head is shaped in order that a claw is cast on the front end of the strand. This provides engagement when the dummy bar and strand are traveling through the strand roller support, and then allows the dummy bar head to be lifted out of engagement.

2.2.5 Strand support systems and secondary cooling

The partly solidified shell as it emerges from the mould is in the region of 10-25mm thick (depending on casting speed) with a surface temperature of around 1000℃ increasing to the solidus temperature (about 1500℃) at the solid/liquid interface. It is subject to the ferrostatic pressure of the liquid steel and would consequently quickly bulge outwards without constraint.

This thin shell, as it emerges from the mould, requires both continual cooling and mechanical support. Secondary cooling sprays are used to control the cooling but the strand support structure, being water-cooled for protection, also extracts heat from the strand. Radiation also contributes to the total heat transfer. The design and operation of the secondary cooling system is dependent on the type and design of the strand support system which in turn depends on the section size and shape being cast. The details of the support equipment for various machines will first be described.

2.2.5.1 Strand support system details for various machine types

The strand support systems vary considerably between those required for billet, bloom and slab casters. For small square sections such as billets the restraining influence of the billet corners are sufficient to prevent shell bulging apart from the region just below the mould. In this case the mould foot rollers combined with support rollers on each side for the first meter or so may be adequate support. This gives more scope further down the strand for more uniform cooling from sprays. However, some billet casters, operating at lower casting speeds and producing section sizes less than about 130mm square or rounds with diameters less than about 150mm have no containment support other than the foot rolls attached to the mould. Any rolls in such machines are usually just to guide the strand and to re-thread the dummy bar. For higher casting speeds for billet casting more support rollers may be required. In any such event the alignment of these rolls with each other and the mould exit is quite important.

The mould length is usually between 700mm and 900mm long but for some billet machines casting at higher speeds a mould extension device is sometimes used. This consists of four spring loaded plates with cooling being provided through orifices in the plates. This mould together with the extension is termed the "Multi Stage (MS) Mould".

For larger billet casters and bloom casting there is an increased propensity for bulging when the shell is still hot and thin and consequently support rolls have to extend further down the strand. Typical support systems for a billet machine and a bloom machine are given in Fig. 2-9.

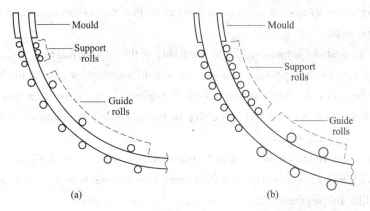

Fig. 2-9 Extent of support rollers for typical billet and bloom casters

For slab machines the bulging of the broad faces extend to the point where solidification is complete and invariably strand support of the wide faces extends the full length of the machines. The latter part of the machine requires rollers for strand withdrawal. Since slab machines are the most complex by both the extent of the support, and the bulging forces involved, the detailed description of the design and operation of strand support systems will concentrate on slab machine requirements. It should be noted that the strand support system contributes significantly to the cooling of the strand and these cooling affects will be included in Section 2.2.6 on "Secondary Cooling".

2.2.5.2 Below mould support system

A variety of strand support and cooling systems just below the moulds in slab machines have or are currently being used. These are: rollers, grids, cooling plates, walking beams. Walking beams proved to be mechanically too complex whilst cooling plates generated too much friction.

The aim is to obtain uniform cooling with minimum friction whilst maintaining accurate support geometry. Today rollers and grids are in most common use with rollers providing the system with least friction between strand and support system.

The secondary cooling arrangement just below the mould very much depends on the strand support system used. For example, with rollers flat sprays are used because of the small gap between the rollers. For grids, however, full cone sprays are used and aligned to direct the cooling water into the rectangular apertures in the grid. With cooling plates, the water is directed through a matrix of small holes and the resulting water film between the plate and the strand provides the cooling.

2.2.5.3 Main strand support systems

In a continuous slab caster, the main support systems are generally composed of segments containing between three to six pairs of rolls with the ability to rapidly exchange the whole segment. The segment frames are clamped together by hydraulic cylinders and the roll gaps are preset using chocks and shims.

The secondary water sprays are aligned on headers so that the solidifying strand is cooled in the gaps between the rolls.

The segment as a whole is fixed rigidly to the frame of the casting machine and the inner radius rolls can be adjusted by the hydraulic cylinders to enable a change of casting thickness(by selection of thicker chocks) or for fully opening which is required in the case of an over cooled slab in the machine which has to be removed by cutting or for scheduled maintenance of the segments in situ.

It is necessary to have the facility to rapidly exchange the segments and Fig. 2-10 shows a schematic diagram of how each segment can be withdrawn from the machine by way of guide rails along which a crane lifts the segments from the machine. In some machines the segments are removed horizontally sideways prior to lifting them out by use of a special crane.

Complex finite element models have been developed to predict the degree of bulging of the solidifying shell both between adjacent roll contacts and when a roll is misaligned with respect to the adjacent rolls. Such models are used to design the optimum diameter and pitch of the support rolls. The pitch has to be such that there is insignificant bulging between the roller contacts and the rolls need to be of such a diameter that minimizes the degree of roll bending due to the ferrostatic force generated by the liquid core and the thermal stresses due to non-symmetrical heating of the rolls.

Up to about 1980 most slab machines used single piece rolls but over the last decade there has

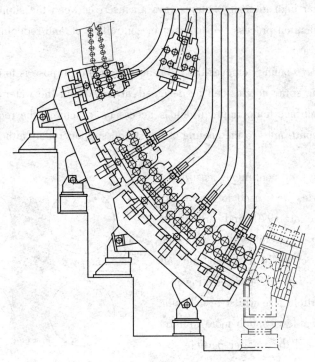

Fig. 2-10 Removal of segments via vertical guide rails

been a significant increase in the application of "divided" or "split" rolls. Single piece rolls extend to over the full width of the strand and are supported by bearings at each end of the roll. With the advent of improved bearing technology (cooling and lubrication in a hot environment) most new wide slab casters and many which have been rebuilt now contain divided rollers. Divided rollers consist of shorter lengths of roller barrels supported part way across the strand by "central" bearings. This allows greater scope to reduce roll diameters and pitches whilst maintaining rigidity and hence roll gap geometry.

Much work has been done to evaluate the performance of various roller designs and details of their behavior as a function of design and other operating parameters are more fully discussed later.

2.2.6 Secondary cooling

Secondary cooling and the containment/withdrawal system extends from the bottom of the mould through complete solidification of the strand to the cutoff operations. The total secondary cooling is a combination of several components which are: cooling due to radiation, cooling due to the water sprays both by the evaporation of the spray water droplets on the slab surface and by the deflected water which accumulates in the entry nip between the rolls. Cooling by conduction to the rolls.

In this section details will concentrate on the water sprays themselves but the design and operation of these sprays are very much dictated by the strand support design and as such the individual effects of the sprays on strand solidification cannot always readily be separated.

As described earlier high intensity water sprays are used between the support rollers to further accelerate the solidification process and to assist in controlling of, and reducing fluctuations in, the strand surface temperatures.

The secondary spray cooling achieves the following: The main purpose is to extract heat from the solidifying strand. The spray nozzles can be designed, arranged and the water flow-rates controlled to give an optimum surface temperature which is necessary to achieve the required surface quality. The spray water contributes to the cooling of the strand support rollers although these are all internally cooled.

In the earlier days of continuous casting of steel water only nozzles were used for secondary cooling but during the late 1970s and early 1980s air-mist sprays were introduced on a wide scale. These consist of both a water and air supply to a nozzle at high pressure resulting in a much finer water particle size whilst also having a wide angle. This enables a much more uniform application of water and the smaller particle size has the advantage of increasing the heat transfer coefficients. Fig. 2-11 shows the two systems.

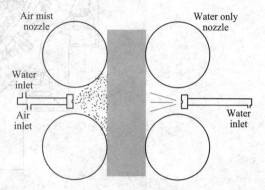

Fig. 2-11 Arrangement of water only and air mist spray systems

The water spray pattern impinging on the strand surface should cover as wide an area as possible but this is often made difficult by the presence of the strand support system. Full cone nozzles are able to cover a large round or square impact area whilst flat spray nozzles can cover a wide impact area across the strand but only a small distance in the direction of casting when used to direct water between adjacent rolls. In billet casters, full cone nozzles are predominately used mounted on header pipes which are installed vertically along each face of the billet strand. The location of support rolls in the upper part of bloom casters and for the whole length of slab casters invariably means that flat spray nozzles have to be used. The length of the entire spray section varies between 0.5m and 6.0m in the case of billet and small bloom casters and can extend up to 20 meters in high speed slab casters. The secondary cooling system is divided into a number of independently controllable zones down the length of the machines. The spray water supply systems are quite independent of both the mould cooling water and the "closed" water system to cool the rolls and bearings and other machine elements.

Where air mist cooling is employed, atomization is by high pressure compressed air acting as the carrier gas. The steam generated is extracted from the spray chamber by large fans. The non-vaporized water which may contain scale and grease is returned down a flume beneath the caster to the water cooling and cleaning plant.

2.2.6.1 Spray cooling with water only

In secondary cooling with water alone, the atomization of the water occurs at the nozzle by virtue of the water supply alone, without additional assistance from other media. In slab casters, the number of horizontal trajectory nozzles located between the rolls determines the system nomenclature. A single-nozzle system denotes the arrangement of one nozzle (occasionally two) which produces a wide-angle spray (up to 120°) at each inter-roll space (spray zone); the multi-nozzle system involves the grouping of many nozzles with a small spray angle at each spray zone. Fig. 2-12 shows these alternate nozzle system arrangements.

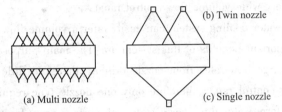

Fig. 2-12   Alternative nozzle sytem arrangements

The single-nozzle system is currently well suited to the majority of the usual slab grades and sizes produced. It began to replace the multi-nozzle system around the mid-1960s because the small nozzle orifices of the latter tended to become clogged very easily. In the meantime, the multi-nozzle system has been revived for certain casters for sheet and sensitive grades, with high spray water flux in conjunction with high casting speeds. The water employed in such systems must have only a minimal content of suspended particles.

The advantages of the single-nozzle system are obvious: fewer nozzles, simpler supply system and easier to maintain. As the single nozzle is installed further away from the strand, it is better protected. Another important benefit in wide-angle single nozzles lies in their relatively high flow capacity (same volume of water with fewer nozzles = greater throughput per nozzle) and hence a larger outlet bore. The outlet bore determines the capacity range of a nozzle, and the flow-rate is controlled within this range by the water pressure. However, large changes in pressure also alter the spray angle, and if the pressure becomes too low, the spray angle collapses and the water flows out of the nozzle orifice without the desired spray effect. The lower pressure limit is generally considered to be 0.05-0.1MPa.

A disadvantage common to all spray nozzles in water-only systems is their comparatively narrow volume flow control range which, given the usual operating pressure encountered in continuous casting plant of 0.1-0.8MPa (at the nozzle tip), is only 1 : 3.5 on average.

In continuous casters in which slabs of various steel grades have to be cast over a very wide range of casting speeds, this limited control range of the nozzles in water only cooling systems may render the installation of two separate spray systems necessary in order to produce the necessary range in water flux. Such systems feature two nozzles of different ratings arranged side-by-side at each cooling zone, and depending on the required water flux, either the smaller, the larger or both nozzles together are employed. Dual systems of this kind are, of course, more expensive and complex.

2.2.6.2 Spray cooling with water and air(air mist)

In water-air mist spray cooling systems, the cooling water is mixed with compressed air in a mixing chamber ahead of the nozzle, and the mixture emerges from the nozzle as a finely atomized, high-impulse, wide-angled spray. This type of spray cooling is particularly suitable for high-grade steels which are susceptible to cracking. Its more important advantages include a particularly uniform cooling pattern and a very wide volume flow control range.

A combined air and water cooling system can easily offer a volume flow control range of 1 : 12 and more. The most important benefits of this system are: The main purpose is to extract heat from the solidifying strand. Large flow-rates from nozzle orifices, therefore little danger of nozzle clogging. Large volume flow control range, therefore only one nozzle type required for all steel grades and casting speeds. Uniform water flux over a wide slab surface area(from roll line to roll line), therefore reduced danger of local over-cooling of the strand surface for a given overall rate of heat extraction. Formation of extremely fine water droplets for optimum cooling effect. Efficient vaporization of the fine droplets results in less water accumulation ahead of the roll nip.

2.2.6.3 Roller design and performance

The design of the support rollers in continuous casting machines for slab production is a compromise of several factors. For slab casting machines installed before 1980 the majority of the support rollers were a single roll with support bearing at each end. In the early 1980s with the advent of the development of bearing technology to resist the adverse environmental conditions in the machine, two or three piece rollers were used.

All rollers and bearings need to be water cooled and apart from some of the smaller rolls in the upper part of the machines (where high secondary water flow-rates are used) all rollers are internally cooled. However, there are several different designs of rollers and the internal cooling efficiency can vary from one design to another. The main requirements of support rolls are: The diameters and pitches should be such that the inter-roll bulging of the strand should be minimized. This in turn depends on the degree of secondary cooling(i.e. the strand temperature), the casting speed (primarily determines shell thickness), the distance down the strand, and the grade of steel. The creep properties of steel can vary significantly depending on steel grade. On a 12m radius machine the ferrostatic pressure at the tangent point is $86t/m^2$ so the force on the solidifying skin is quite large. The degree of bulging is also time dependent and therefore the time taken for a particular element of the solidifying shell to pass from one roll to the next is related to casting speed. Geometrically the rolls should remain stable. If the rolls were too small in diameter and maybe 2 meters long(a typical slab single roll length) then the rolls would bend due to: the ferrostatic force, the thermal stresses since the rolls have an asymmetrical temperature distribution during operation. During a strand stoppage the asymmetric temperature is magnified considerably.

The water cooled support rolls themselves can extract a significant amount of heat from the solidifying strand and the amount of heat extracted depends on the roll design. The various types of roll designs and roll cooling methods are illustrated in Fig. 2-13 which shows the main roll design and cooling methods. Examples are for single piece rolls but many of the principles also apply to divided rolls.

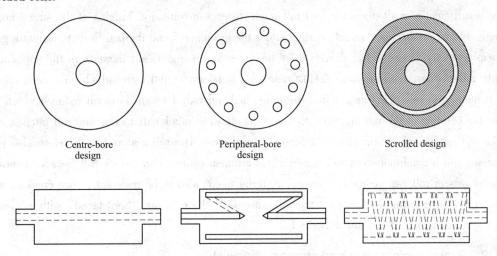

Fig. 2-13 The different types of internal roll cooling

Because the cooling channels of the peripheral-bore design and the scrolled design are near the surface the roll surface is kept colder. These are commonly called cold roll designs whilst the centrally bored cooling is termed a "hot" roll design. The cold roll designs extract significantly more heat from the strand than does the hot roll design. However, the cold roll designs are more stable and much less prone to permanent bending when the strand stops and the roll bends due to grossly asymmetric temperature distribution leading to severe thermal stresses. If a roll becomes permanently bent to a degree of greater than about 1mm at the center this can lead to poor internal quality.

Much work has been done on evaluating roll performance both in terms of geometrical stability and heat extraction capability. It is interesting to note that the amount of spray water used affects the heat extracted by the roll. With no spray water entering the roll gap the heat extraction is 44kW/m and 26.5kW/m respectively. (These values are the kW per meter length of roll).

Data have also been obtained on the geometrical stability of the various types of roll design. Bulgemeters have been used to measure both roll behavior and the bulging of the strand. These bulgemeters consisted of linear displacement transducers (LDT) on the end of units which were rigidly fixed in the machine with the LDTs resting on the back of the rolls or the strand surface as appropriate.

Three such bulgemeters at any single location in the strand are used, two on adjacent rolls and one on the strand between the two rolls. These instruments can be left in the strand over long periods and the behavior of the rolls and strand have been investigated for many events such as strand stoppages or slowdowns and for various secondary cooling conditions in casting different steel grades.

Such deviations of roll geometry need to be avoided since these lead to unacceptable surface and internal quality. This is described in greater detail later.

All the work just described was carried out on single piece rolls which have to compromise between a sufficiently small diameter(and roll pitch) to prevent inter-roll bulging of the strand and a sufficiently large diameter to avoid bending under the mechanical and thermal loads to maintain good roll gap geometry. Over the last decade there has been a very significant increase in the application of split rolls as described previously. Most new wide slab casters and many which have been rebuilt now contain split rolls. This means that the individual roll barrel length is much reduced which reduces the bending of the roll significantly and thus allowing smaller diameters and roll pitches.

The roll gap geometry can also be affected by roll wear. The roll material is therefore also very important and a combination of roll material and efficient cooling can reduce roll wear as a serious cause of loss of roll gap geometry. The roll material needs also to be resistant to fire cracking and stress corrosion cracking and to meet these requirements the rolls are "hard faced" with a layer of metal comprising 12wt% Cr and 88wt% Fe.

2.2.7 Strand straightening and strand withdrawal

For casting machines, where the strand is either cast in a curved mould or is bent into a curved position below the mould, the strand requires to be straightened before it can be discharged horizontally. The design of the straightener(or the bending zone where the strand is curved after being cast in a vertical mould) is dependent on machine radius, section size, steel grades to be cast and other casting parameters. Details will be described in Section 2.2.7.1 below.

Additionally, sufficient power and traction need to be imparted to the strand to enable withdrawal to be reliable and consistent.

2.2.7.1 Strand straightening

As indicated previously, the curved strand needs to be straightened to achieve horizontal discharge. The design of the straightening unit depends on several factors and it is important to ensure that any stresses caused by the strains imposed due to straightening are smaller than the inherent strength of the material.

The strain distortion across the fully or partially solidified strand can be determined from standard beam bending theory but due to the temperatures involved creep occurs and hence to design for the overall strains required to straighten the strand the strain rate is also an important consideration. The strain distribution across the strand also depends on whether the strand is completely solid or whether a liquid core still exists. In modern machines requiring higher throughput a liquid core usually exists during straightening. The two situations will be dealt with separately.

A Strand completely solidified

The strain distribution in this case depends entirely on the initial curvature and strand thickness and is shown in Fig. 2-14.

The surface strain is

$$\varepsilon_s = \frac{b}{2R} \times 100\% \qquad (2-8)$$

this being a tensile strain on the top surface and a compressive strain on the bottom surface. The strain rates can be reduced by applying the required strain over more than one unbending point or even continuously straightening over a given length of strand. These systems will be described later.

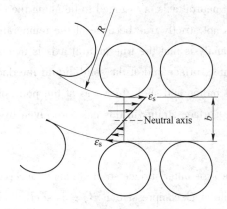

Fig. 2-14 Stain distribution across the solidified strand during single point straightening

B Straightening with a liquid core

In this case both the upper and lower solidified shell is considered as separate beams but the calculated strains can depend on the constraining influences of the solidified edges. These can be significant at low aspect ratios, when the solidified shell has reached a significant thickness and the shape of the shell has been influenced by the two dimensional heat transfer. Two approaches are therefore adopted. These are termed the "Soft Box" and "Hard Box" approach respectively.

"Soft Box" approach: The strand is considered to be a soft box when the upper and lower solidified shells deform independently of each other i. e. there is no restraining influence of the solid edges. This is the situation in the case of a slab where the aspect ratio is high and the shell thickness small compared to the slab width. Fig. 2-15 shows the strain distribution occurring in the solidifying shell due to straightening at the tangent point.

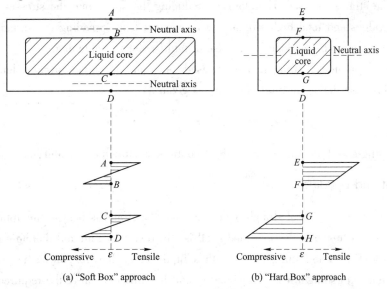

Fig. 2-15 Strain distribution in solidifying shell

The neutral axis is assumed to be along the centerline of both the upper and lower shell although this is not strictly true because of the temperature gradient. It has been shown by using finite element analysis that the true neutral axis is nearer the cold surface. There are tensile strains both at top outer surface and at the solid/liquid interface of the lower shell. These strains are a function of strand radius and shell thickness at the point of straightening.

The surface strains in this case are given by:

$$\varepsilon_s = \varepsilon_i = \frac{t}{2R} \times 100\% \tag{2-9}$$

Where ε_s is outer surface strains (tensile at A; compressive at D); ε_i is solid/liquid interface strains (tensile at C; compressive at B); t is shell thickness, m; R is machine radius, m.

"Hard Box" approach: In this case the bending is primarily influenced by the stiffness of the solidified edges and the neutral axis in this case is assumed to be along the section mid thickness and the surface strains are similar to the situation where the strand is totally solid i. e.

$$\varepsilon_s = \frac{b}{2R} \times 100\% \tag{2-10}$$

The solid/liquid interface strains in this case are given by

$$\varepsilon_i = \frac{b - 2t}{2R} \times 100\% \tag{2-11}$$

It has been demonstrated that the soft box approach is appropriate for slabs or large blooms with a high aspect ratio. The hard box approach is only applicable to billet and small bloom sections.

As indicated earlier the strain rate often determines whether a crack defect (either internal or on the surface) will occur. The inherent strength of the steel particularly at the solid/liquid interface is very low at the temperatures involved but at these temperatures creep rapidly reduces stresses resulting from the strains imposed. Therefore, by reducing the strain rate the stresses can be maintained at low values and total high strains can be achieved by spreading the straightening over a length of the machine. This is done by the use of multi point straightening.

In the limit continuous straightening is used on some machines over a length L of the machine. In this case the strain rate (per min) is given as:

$$\dot{\varepsilon}_s = \frac{\varepsilon_s v}{L} \tag{2-12}$$

where v is casting speed, m/min; L is length of continuous straightening unit, m.

2.2.7.2 Strand bending

In the situation where a vertical straight mould is used the strand is bent to the appropriate radius below the mould. In this case the solidified shell is still relatively thin and therefore the strains are usually not as high as when straightening with a liquid core. However, the same principles apply and many casters with straight moulds use multi point bending to achieve the required radius whilst reducing the strain rates to avoid internal defects. In such cases misalignment of the bending rolls again requires to be minimized to reduce misalignment strains.

2.2.7.3 Withdrawal units

The strand needs to be withdrawn from the machine under constant and controlled conditions and sufficient power and traction needs to be applied to achieve this. The withdrawal force has to be sufficient to overcome the frictional forces acting on the strand. These can arise due to: strand friction in the mould, friction of the support rolls in their bearings resulting from their operating loads, rolling friction owing to strand bulging between the rolls.

It should also be noted that the dead weight of the strand itself acts in favor of reducing the required withdrawal force.

Fig. 2-16 shows examples of withdrawal units for a bloom machine and a slab machine.

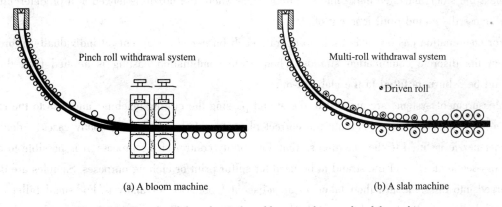

Fig. 2-16 Strand withdrawal unit for a bloom machine and a slab machine

Modern withdrawal units for slab machines are multi roll withdrawal systems, the traction and power being distributed over several roll pairs. The drive roll pairs achieve the correct amount of traction by the use of hydraulic forces slightly in excess of the ferrostatic force at that position. The withdrawal forces occurring in slab machines can only be overcome by the multi roll withdrawal system. Such a system successfully reduces the strand withdrawal force at an early stage, reducing it to a low level as the strand progresses through to caster.

The tensile force just below the mould shows a slight initial decrease owing to the dead weight of this strand. It remains at the relatively low value until the strand reaches the straightening section where it abruptly increases in magnitude. Following complete solidification, the rate of increase eases due to the elimination of ferrostatic forces.

2.2.8 Product discharge and handling

Following complete solidification, the strand needs to be divided into the piece sizes that are forwarded to the mill for subsequent processing. After cut-off, the process changes again from a continuous operation to a batch operation.

This batch operation must be designed so that the cycle time of individual elements of the discharge are faster than the delivery of cut pieces into the system from the cut-off. Otherwise, the continuity of the earlier operation will be interrupted.

The cut-off unit is located as close as possible to the end of the strand support or final solidification. With the exception of small billets or thin slabs, which can be cut by shears, all cutting is undertaken by oxy-fuel gas cutting. Gas cut-off is used, as this is the most cost-effective means of cutting the strand. However, there is a small "kerf" loss due to the cut. This kerf loss is in the order of 3/8-1/2 inch(8-12mm). For small billets or thin slabs, cut-off shears can be applied; these are more expensive, but the yield loss is zero.

The cut-off is achieved by burning oxygen and a fuel gas, typically propane or coke oven gas, in a specially designed nozzle. The oxygen purity usually needs to be at least 99.5%. The nozzle or nozzles are arranged to approach the edge of the strand, heat the edge for a short period in order to preheat the steel, and then move into a cutting cycle where the nozzle is moved at a predetermined rate across the strand until it is cut off.

For combination casting, it is usual to operate each burner of a slab cut-off individually in order to cut the discrete combination strands. When triple combination casting is applied then there should be 3 burners fitted to the slab cut-off.

Measurement systems are applied to the strand passing the cutting machine, and also to the machine position, in order to calculate the correct place to cut the product. In many cases, a double burner nozzle is fitted to the machine so that, for quality control of the process, it is possible to cut a cross-sectional slice of the strand to be used for sulfur print or etching purposes. Samples are discharged into bins and are then taken to an adjacent laboratory for analysis. For small billet and thin-slab casters, shears can be applied. These units are generally too expensive to apply to larger section sizes. The main benefit of shears is that they do not create a yield loss due to cutting.

3 Heat Transfer and Strand Solidification

3.1 Mould Heat Transfer

The heat transfer details, mechanisms and the solidification behavior in the water cooled copper mould are among the most important processes taking place during the continuous casting of steel. It is fundamental that the mould extracts heat from the steel in as uniform a manner as possible with some degree of control. The surface quality of the cast semi is very dependent on mould parameters since this is where the surface is formed and can, therefore, be the source of many surface defects. Uniform heat transfer also helps to avoid breakouts.

Fig. 3-1 shows the temperature distribution between the solidifying steel and the cooling water.

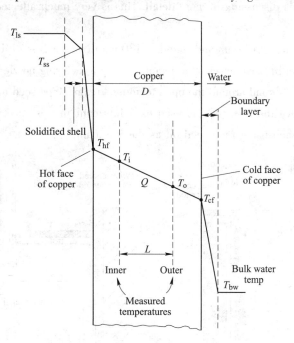

Fig. 3-1 Temperature distribution between steel and cooling water

The heat flux Q is given by:

$$Q = h_{ss}(T_{ss} - T_{hf}) = \frac{K}{D}(T_{hf} - T_{cf}) = \frac{h_{cf}(T_{cf} - T_{bw})Kw}{m^2} \tag{3-1}$$

Where h_{ss} is heat transfer coefficient from the face of the solidifying steel, kW/(m² · K); T_{ss} is temperature of the outer face of the solidifying steel, ℃; T_{hf} is copper "hot face" temperature, ℃; T_{cf} is copper "cold face" temperature, ℃; K is thermal conductivity of copper, kW/(m · K); h_{cf} is heat transfer coefficient of the "cold" copper face, $kW/(m^2 \cdot K)$; T_{bw} is bulk temperature of the cooling water, ℃; D is thickness of copper, m.

From the liquid steel temperature in the mould, there is a temperature drop across the solidifying skin which will be discussed more fully later. The interface between the steel shell and the hot face of the mould wall incorporates the film of lubricant and any gaps which form and this component of the heat transfer represents a major factor governing the heat flux from the steel to the cooling water in the mould. The high conductivity of the mould wall material ensures a small temperature drop across the copper. The cold face of the mould wall can be significantly higher than the bulk cooling water temperature due to the boundary layer which is present in any water cooling channel. This boundary layer, however, can be affected by the cooling water flow conditions in the cooling channel and the temperature drop across the boundary layer can be fairly confidently predicted from well proven heat transfer theory. It is necessary to maintain the cooling water velocities sufficiently high (8m/s) to avoid nucleate boiling.

The interface between steel and copper, the major component to the thermal impedance, is a complex area and needs discussing in more detail. This is very much affected by the type of lubricant used.

In billet casting squares <130mm and rounds <130mm diameter it is difficult to use a refractory submerged entry nozzle. In these cases open teeming using a metering nozzle is practiced but invariably using an inert gas shroud around the open teeming stream. Rape seed oil, fed from small holes in the copper face above the meniscus, is used as a lubricant in this case. Fig. 3-2(a) shows the details in the mould when using rape seed oil as the lubricant.

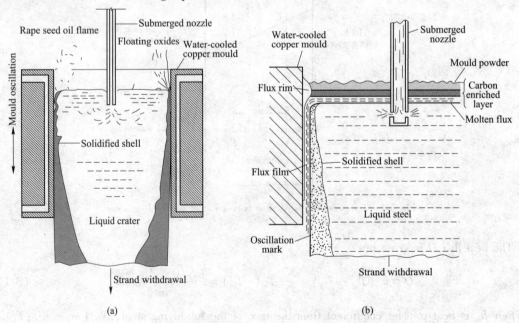

Fig. 3-2 Teeming and mould details for lubrication using
(a) rape seed oid and (b) mould powder to provide a slag

In slab and bloom casting a submerged entry nozzle(SEN) is used together with a synthetic mould powder which forms a fluid slag between the powder and the steel in the mould. Fig. 3-2(b) shows the details in the mould and the interface with the copper when using a submerged entry nozzle and synthetic powder.

The main advantages of using mould powder over rape seed oil are:

(1) A submerged entry nozzle(SEN) is used with mould powder which is a more efficient method of stream shrouding.

(2) It prevents radiative heat losses from the metal surface in the mould and prevents solidification on the surface which can lead to "plating" defects.

(3) The slag formed from the powder absorbs non-metallic inclusions(e. g. Al_2O_3) which float out of the metal pool in the mould.

(4) The slag allows more uniform heat transfer to the copper wall.

The mould powder composition and properties needs to be such that the heat from the liquid steel produces a continuous fluid slag layer of adequate thickness and with a viscosity which enables a continuous flow of slag into the meniscus at the copper wall.

A further fundamental requirement is that the mould is oscillated sinusoidally in such a manner that for a certain percentage of the cycle the mould would be travelling in a downward direction faster than the solidifying shell.

Fig. 3-3 shows the oscillation cycle and that part of the cycle where the mould travels downwards faster than the strand. This is called the negative strip time, or heal time, and is chosen as a compromise between lubrication(and hence friction) and the maintenance of uniform heat transfer. This will be discussed in much more detail later.

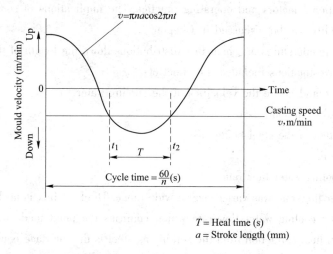

T = Heal time (s)
a = Stroke length (mm)

Fig. 3-3 Oscillation cycle showing negative strip time

The interactions between the mould oscillation, mould slag feeding and variations in mould metal levels are quite complex and several computer models have been developed to determine the mould powder consumption rate and the solidification characteristics at the meniscus as a function of mould oscillation, mould level and powder slag properties. This will be discussed further in Section 3.4.2.

All these factors determine the slag film thickness which in turn determines the thermal impedance of the interface. Additionally, the gap between the solidifying steel and the copper wall is affected by the surface temperature and shell contraction, which can cause air gaps to form which may depend on section size and shape. Fig. 3-4 shows for various strand cross-sections the formation of an air gap between the strand shell and the mould wall such as occurs below the meniscus level.

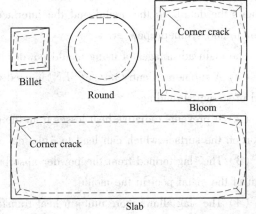

Fig. 3-4 Gap formation and change in cross-section resulting from shrinkage in the mould

These gaps can also vary down the length of the mould usually increasing from below meniscus level. This is counteracted by a three-dimensional taper for billet and bloom cross-sections. In the case of slab moulds only the narrow faces follow the shrinkage in the cross-section and only the end plates are consequently tapered. Due to bulging no gaps form along the broad faces for slabs and the broad faces are set parallel to each other.

Much work has been carried out using thermocouples embedded in the mould copper plates to measure the heat flux through the mould plates. This work has generally concentrated on the interrelation of the heat flux, heat flux distribution, mould wall temperatures, type of mould lubricants used, steel compositional factors and operating practice. The implications of some of these factors on as-cast steel quality will be discussed in Chapter 3.

Both mould wall temperatures and heat flux distributions down the length of the mould were investigated. These investigations included the effect of:

(1) The flow-rate and hence the velocities of the cooling water.

(2) Type of mould lubricant used.

(3) Carbon content of the steel being cast.

(4) Casting speed.

(1) Effect of cooling water flow rate.

The cooling water flow-rate was varied over a wide range. The heat flux is fairly constant for this wide variation in the cooling water flow-rate which confirms the point made that the over-riding controlling factor on heat extraction from the solidifying steel is the interface boundary between the steel and the hot face of the copper mould. The effect of the boundary layer can be seen to have driven the copper temperatures higher for a lower water flow rate.

(2) Effect of mould lubrication.

Fig. 3-5 shows the effect of various mould lubricants on the heat flux distribution down the mould and on "hot" face copper temperatures. These distributions are shown for two mould casting powders and for rape seed oil. As can be seen from these results, the heat fluxes are considerably higher for rape seed oil and it is worth noting in particular the very increased heat flux in the meniscus region.

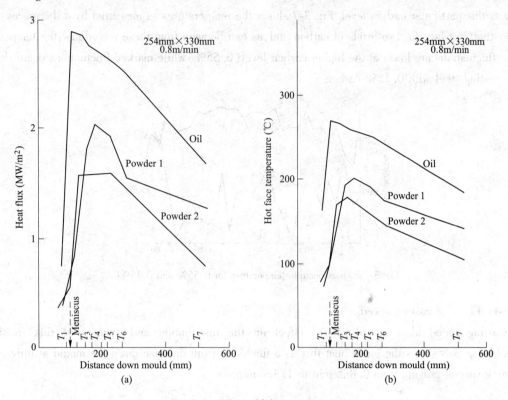

Fig. 3-5 Effect of lubricant type on
(a) heat fluxes and (b) hot face temperature

(3) Effect of carbon content.

The effect of steel composition and particularly carbon content on the overall mould heat transfer has been reported from several sources. Fig. 3-6 shows the measured average heat flux in the mould over the carbon range of 0.02% to 1.6%. The effect of carbon content on heat transfer leads to some quality problems being more acute within the carbon range 0.06% to 0.14% (the peritectic range).

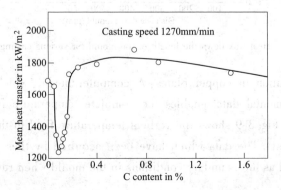

Fig. 3-6 Effect of carbon content on mould heat flux

Irregular shell thicknesses down the length of the mould have been observed for 0.1% carbon steels. It was proposed that this irregularity in shell thickness and in non-uniform heat transfer is caused by the γ to δ phase transformation and associated volume changes and shrinkages which occur at this particular carbon level. Fig. 3-7 shows the temperatures as measured by a thermocouple near the front face for two levels of carbon and, as can be seen from these recordings, the temperature fluctuations are lower at the higher carbon level (0.55%) while marked fluctuations occur during casting steel with 0.15% carbon.

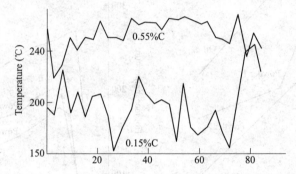

Fig. 3-7 Thermocouple temperature for 0.55% and 0.15%C

(4) Effect of casting speed.

Casting speed also has a marked effect on the distribution and mean heat flux in the mould. Fig. 3-8 shows the mean heat flux as a function of the distance down the mould at different casting speeds ranging from 0.8m/min to 1.3m/min.

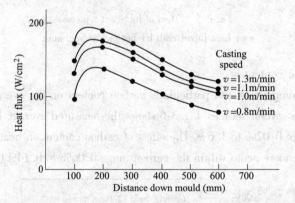

Fig. 3-8 Heat flux down the length of the mould for various casting speeds

Temperature distribution in copper plates: A computer model, which has been calibrated with the above experimental data, enables the complete temperature field within the mould walls to be calculated. Fig. 3-9 shows the vertical temperature field in the mould wall material along with measured data. The data which have been acquired by these many extensive plant measurements are used as the boundary conditions in the mould when running the strand solidification.

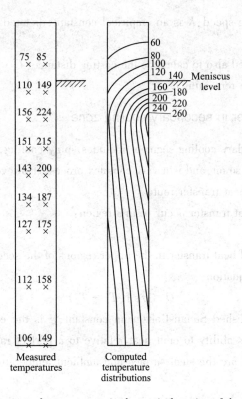

Fig. 3-9 Computed and measured temperatures in the vertical section of the 254mm copper end plate

3.2 Secondary Cooling Zone Heat Transfer

Typically, the secondary cooling system is comprised of a series of zones, each zone responsible for a segment of controlled cooling of the solidifying strand as it progresses through the machine. The sprayed medium is either water or a combination of air and water.

3.2.1 Solidification processing

As the shell thickness increases toward the end of this so-called secondary cooling zone, the supporting rolls grow larger and are spaced farther apart. The secondary zone is often also called the metallurgical length, because this is where the strand solidifies and the cast structure develops. Depending on the strand's cross section and the casting speed, it can be 10 to 40 meters long. The flow of water to the many nozzles in the various sections is often computer-controlled and automatically adjusted as casting conditions change.

After the strand passes through the last pair of support rolls, it enters the run-out table and is cut, while moving, by one or two oxyacetylene torches.

Shell growth can be reliably predicted from Fick's Law:

$$L = v\left(\frac{D}{K}\right)^2 \tag{3-2}$$

Where, D is the Shell thickness; L is the cast distance from mould steel meniscus, where solidification begins; v is the casting speed; K is an empirical constant, dictated mostly from cast grade and machine design.

This equation can be used also to calculate the casting distance (L) where the product is fully-solidified (i.e. no liquid core remaining).

3.2.2 The heat transfer in secondary cooling zone

The heat transfer in secondary cooling segment includes spray cooling, natural convection, thermal radiation, roll contact, and so on, and is a very complex process. However, spray cooling is the most significant process in the heat transfer route.

Three basic forms of heat transfer occur in this region:

(1) Radiation.

The predominant form of heat transfer in the upper regions of the secondary cooling chamber, described by the following equation:

$$Q = \sigma \varepsilon A (T_s^4 - T_a^4) \qquad (3-3)$$

Where σ is the well-established Stefan-Boltzmann constant; ε is the emissivity constant, typically 0.8 (a measure of a body's ability to emit heat relative to a perfect radiator or black body); A is the surface area; T_s and T_a are the steel surface and ambient temperatures respectively.

(2) Conduction.

As the product passes through the rolls, heat is transferred through the shell as conduction and also through the thickness of the rolls, as a result of the associated contact. This form of heat transfer is described by the Fourier Law:

$$Q = \frac{kA(T_i - T_o)}{\Delta x} \qquad (3-4)$$

For conductive heat transfer through the steel shell, k is the shell's thermal conductivity, whereas A and Δx are the cross-sectional area and thickness of the steel shell, respectively, through which heat is transferred. T_i and T_o are the shell's inner and outer surface temperatures, respectively. As shown in Fig. 3-10, this form of heat transfer also occurs through the containment rolls.

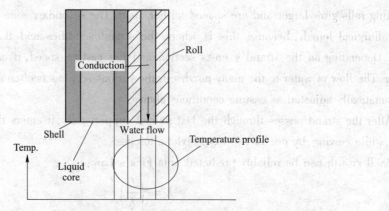

Fig. 3-10 Solidification profile through steel shell & roll

(3) Convection.

This heat transfer mechanism occurs by quickly-moving sprayed water droplets or mist from the spray nozzles, penetrating the steam layer next to the steel surface, which then evaporates. This convective mechanism is described mathematically by Newton's Law of Cooling:

$$q = hA(T_s - T_w) \tag{3-5}$$

Where the coefficient of heat transfer, h a constant, is determined experimentally for selected water fluxes, nozzle types, spray water pressure (and air pressure), and finally steel surface temperature; A is surface area; T_s and T_w are the steel surface and spray water temperatures respectively.

Specifically, the spray chamber (Secondary Cooling) heat transfer serves the following functions.

(1) Enhance and control the rate of solidification and for some casters achieves full solidification in this region.

(2) Strand temperature regulation via spray-water intensity adjustment.

(3) Machine containment cooling.

The containment region is an integral part of the secondary cooling area. A series of retaining rolls contain the strand, extending across opposite strand faces. Edge roll containment may also be required. The focus of this area is to provide strand guidance and containment until the solidifying shell is self-supporting.

In order to avoid compromises in product quality, careful consideration must be made to minimize stresses associated with the roller arrangement and strand unbending. Thus, roll layout, including spacing and roll diameters are carefully selected to minimize between-roll bulging and liquid/solid interface strains.

Strand support requires maintaining strand shape, as the strand itself is a solidifying shell containing a liquid core that possesses bulging ferro static forces from head pressure related to machine height. The area of greatest concern is high up in the machine. Here, the bulging force is relatively small, but the shell is thinner and at its weakest. To compensate for this inherent weakness and avoid shell rupturing and resulting liquid steel breakouts, the roll diameter is small with tight spacing. Just below the mould all four faces are typically supported, with only the broad faces supported at regions lower in the machine.

3.3 Solidification of the Continuous Casting Strand

3.3.1 The solidification process

Solidification in continuous casting technology is initiated in a water-cooled open-ended copper mould. The steel shell which forms in the mould contains a core of liquid steel which gradually solidifies as the strand moves through the caster guided by a large number of roll pairs. The solidification process initiated at meniscus level in the mould is completed in secondary cooling zones using a combination of water spray and radiation cooling. Solidification specialty of CC technology arises from the dynamic nature of the casting process. In particular this relates to:

(1) Handing of very high heat flux in the mould.

(2) Nurturing of the initial thin and fragile solid shell for avoidance of breakout during descent of the strand down the mould.

(3) Designing of casting parameters in tune with the solidification dynamics of the steel grade, for elimination of surface and internal defects in the cast product.

During solidification the solid/liquid interface can be either stable or unstable. For pure metals, the interface velocity is controlled by the diffusion of the latent heat away from the interface. A stable interface is characterized by the rapid conduction of latent heat from the interface through the solid. Any perturbation on the interface will be come up against liquid and will disappear. Therefore, all points along the interface will be move with the same velocity. For an unstable interface, on the other hand, the liquid temperature is below the melting temperature. The liquid is under cooled, resulting in heat flowing away from the interface through the liquid. Thus any perturbation on the interface will come up against colder liquid that enables more heat release from the interface and the perturbation will, therefore, grow faster than the other points on the interface and from a dendrite arm or a side branch.

Unstable solidification is more common during alloy solidification since, for alloys, unstable interface results from species concentration difference near the interface. In pure metals high cooling rates can reduce the temperature of the melt well bellow the melting temperature before solidification process takes place. The under cooling can be as much as 10% to 30% of T_m, where T_m is the absolute melting temperature, depending on the cooling rate and the purity of the melt.

3.3.2 Solidification of the molten steel in mould

The main function of the mould is to establish a solid shell sufficient in strength to contain its liquid core upon entry into the secondary spray cooling zone. Key product elements are shape, shell thickness, uniform shell temperature distribution, defect-free internal and surface quality with minimal porosity, and few non-metallic inclusions.

To understand the phenomena of initial solidification in the mould of a continuous caster it is necessary to understand the broader area of solidification in general. The solidification of a metal alloy is a complicated process which involves, on the macroscopic scale, heat transfer through the solid metal, the liquid metal, and the mould with a continually moving interface between the solid and the liquid as solidification progresses. On the microscopic scale, solidification involves nucleation, heat transfer, interfacial energy, and segregation. The mathematical analysis required for the description of the phenomena is complex and generally requires calculations to be performed by iterative techniques. A basic understanding of solidification, however, can be gleaned from a review of the pertinent theories and equations.

The mould consists of an opened, water-cooled copper box or cylinder. This box is rigidly affixed to a steel support structure and has a series of water channels adjacent which act to cool the mould. Mould oscillation is used to prevent sticking of the mould. To further facilitate lubrication either mould slag or oils are infiltrated in the space between the casting and the mould, where the solidifying slag provide heat transfer control. This whole apparatus is set above a series of rollers for

3.3 Solidification of the Continuous Casting Strand

bending and straightening of the strand; and water jets which cool the strand.

Solidification in the mould is a crucial phenomenon in the determining of the quality of continuous cast product. For the case of a slab caster, the molten steel is introduced into the mould through a bifurcated submerged entry nozzle, and the liquid streams are directed at the narrow faces of the rectangular mould. On the top of the steel is a liquid mould slag which results from the fusion of the added mould powder. A mould slag is a mixture of liquid oxides and other compounds, and contains a mixture of alumina, silica, lime, soda, and certain fluorides. The mould slag acts to protect the steel from oxidation, as well as lubricate the strand and control heat transfer. The steel contacts the mould flux which has been chilled by contact with the copper mould surface, quickly solidifies and the initial shell of the casting. As casting continues, the shell is drawn downwards and grows. Moulds are generally around one meter in length, and the steel has a residence time which varies from one minute in bloom casting to around 20 seconds in the case of slab casting. This is not enough time for total solidification of the strand, but it is enough time for the development of a strong enough shell for containment in the secondary cooling zones of the caster. If the shell is not sufficiently thick, for whatever reason, breakouts where liquid steel drains out through the cast shell can occur. After the slab exits the mould it is further cooled by spraying of water directly on its surface, and by heat transfer into the rolls. A schematic of the mould and the solidification conditions within it are shown in Fig. 3-11.

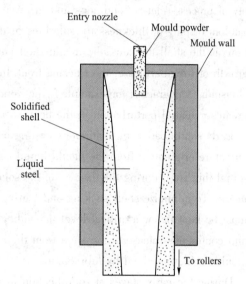

Fig. 3-11 Schematic of continuous caster mould

During the casting process, an initially liquid mass of material is caused to solidify by reducing its temperature below the melting point by heat transfer from its surfaces, which are in contact with a surrounding mould. The process therefore generally starts by the formation of a solidified shell in contact with the mould and surrounding the remaining liquid mass.

If solidification is interrupted, periodic thickness non-uniformities at the freezing front are often observed that can have wavelengths of the order of several centimeters. If solidification is allowed to proceed, however, the non-uniformities tend to die out as the freezing front morphology becomes less dependent upon the mould-shell interface due to the ever-thickening shell. Such microstructures are detrimental to subsequent forming processes and have linked to severe ingot cracking. Experimental observations of thermo-mechanical growth instability in casting processes have been reported by Cisse et al. It has been suggested that patterned mould surface geometries may help to promote the uniform growth of a casting and hence deter the onset of the observed shell thickness non-uniformities.

Murakami et al. proposed that periodic grooves in the mould surface, which led to gaps of a controlled size along the mould-shell interface due to imperfect wetting of the molten metal, resulted in a number of important improvements. Perhaps, the most significant improvements were more uniform contact along the mould-shell interface, and a reduction in crack nucleation in the ingot due to slower, but more uniform heat extraction. To test this hypothesis, the immersion tests were repeated for aluminum alloys using casting moulds with machined grooves. A key parameter that was investigated was the groove pitch or wavelength and their impact on the shell thickness non uniformity. It was suggested that there was a possibility of a wavelength selection process where in the system "picked-off" a mould surface wavelength or band of wavelengths such that the shell grew with greater uniformity.

For a smooth mould surface (or at least one with no prominent periodicity), the perturbations in heat extraction result from stochastic variations in the mould-shell interface heat flux due to a variety of process-related conditions and material properties/metallurgical transformations. An equally random display of thickness irregularities in the shell results during the early stages of solidification: as the shell thickens, the mould-shell boundary conditions have a diminishing impact on the growth of irregularities at the freezing front. In the idealized case of a mould surface with a purely sinusoidal topography (for example), the controlling factor is the topography geometry, since this creates a spatial perturbation in the heat extraction profile.

Early solidification in continuous casting occurs in the form of partial freezing of the meniscus curvature originating from the mould liquid contact point. Prevention of sticking and tearing of this initial thin shell during the descent of the solidifying strand is one of the major functions of the CC mould. To minimize shell sticking and tearing, friction between the strand surface and mould wall must be kept below a critical level depending upon the shell strength. Minimization of the friction and continuous release of the shell from the mould have been achieved through the introduction of mould oscillation aided by lubrication.

During the early stages of solidification in near net shape processes, the total resistance to heat transfer from melt to the mould is mainly governed by the thermal resistance of the metal/mould interface, as opposed the conventional continuous casting, where the resistances of the solidifying shell and of the lubricant film are also significant. In the twin-roll continuous casting operations, the molten metal is solidified directly on a substrate (two water cooled rolls) without a lubricant.

The maximum casting speed of a strand is limited by the growth and the strength of the solidifying shell within the water-cooled copper mould.

As in ingot casting, for the continuous casting process, the approximation formula (3-6), that the thickness, S of the solidified strand shell is proportion to the square root of the solidification time. t is also valid:

$$S = K\sqrt{t} \tag{3-6}$$

The factor K has an "average" value of 26, where S is in mm and t in min. In practical and for lay-out calculation, values in the range 23 to 32 are used to compensate in a simple manner for different influences of, for example, steel grades, strand sizes, spray water ratings, etc.

The total solidification time is thus determined by K and, for a rectangular strand cross section, by half the strand thickness. A thick strand needs a much longer solidification time than a thin one, time increasing or, for a given machine length, casting speed decreasing with the square of the thickness. This has a direct effect on machine design because-at least in the case of slabs-the strand has to be protected against bulging as long support. In other words, strand thickness and casting speed determine the necessary machine length or "metallurgical length".

In the steel manufacturing by continuous casting, the moulding process is a critical path because steel quality is significantly influenced here. When the molten steel is poured from a tundish into the mould, solidification takes place along the mould wall and forms a shell. The solidified steel shell continuously grows in thickness and becomes solid slab or fillet. The moulding process is also critical in the mathematical modeling point of view since fluid coexists together with solid and solidification is in process. The molten steel flow and the temperature distribution determine the solidification paten and further microstructure of the solid steel. Thus, the solidification process would make a significant effect on the steel quality and productivity. An appropriate flow control in the moulding process is essential for quality steel production.

3.4 Modeling of Continuous Casting

The high cost of empirical investigation in an operating steel plant makes it prudent to use all available tools in designing, troubleshooting and optimizing the process. Physical modeling, such as using water to simulate molten steel, enables significant insights into the flow behavior of liquid steel processes.

The complexity of the continuous casting process and the phenomena which govern it make it difficult to model. However, with the increasing power of computer hardware and software, mathematical modeling is becoming an important tool to understand all aspects of the process.

3.4.1 Physical models

Previous understanding of fluid flow in continuous casting has come about mainly through experiments using physical water models. This technique is a useful way to test and understand the effects of new configurations before implementing them in the process.

Construction of a physical model is based on satisfying certain similitude criteria between the model and actual process by matching both the geometry and the force balances that govern the important phenomena of interest. To reproduce the molten steel flow pattern with a water model, all of the ratios between the dominant forces must be the same in both systems. This ensures that velocity ratios between the model and the steel process are the same at every location. It shows some of the important force ratios in continuous casting flows, which define dimensionless groups. The size of a dimensionless group indicates the relative importance of two forces. Very small or very large groups can be ignored, but all dimensionless groups of intermediate size in the steel process must be matched in the physical model.

An appropriate geometry scale and fluid must be chosen to achieve these matches. It is fortunate that water and steel have very similar kinematic viscosities (μ/ρ). Thus, Reynolds and Froude numbers can be matched simultaneously by constructing a full-scale water model. Satisfying these two criteria is sufficient to achieve reasonable accuracy in modeling isothermal single-phase flow systems, such as the continuous casting nozzle and mould, which has been done with great success. Actually, a water model of any geometric scale produces reasonable results for most of these flow systems, so long as the velocities in both systems are high enough to produce fully turbulent flow and very high Reynolds numbers. Because flow through the tundish and mould nozzles are gravity driven, the Froude number is usually satisfied in any water model of these systems where the hydraulic heads and geometries are all scaled by the same amount.

Physical models sometimes must satisfy heat similitude criteria. In physical flow models of steady flow in ladles and tundishes, for example, thermal buoyancy is large relative to the dominant inertial-driven flow, as indicated by the size of the modified Froude number (Froude*), which therefore must be kept the same in the model as in the steel system. In ladles, where velocities are difficult to estimate, it is convenient to examine the square of the Reynolds number divided by the modified Froude number, which is called the Grashof number. Inertia is dominant in the mould, so thermal buoyancy can be ignored there. The relative magnitude of the thermal buoyancy forces can be matched in a full-scale hot water model, for example, by controlling temperatures and heat losses such that $\beta \Delta T$ is the same in both model and caster. This is not easy, however, as the phenomena that govern heat losses depend on properties such as the fluid conductivity and specific heat and the vessel wall conductivity, which are different in the model and the steel vessel. In other systems, such as those involving low velocities, transients or solidification, simultaneously satisfying the many other similitude criteria important for heat transfer is virtually impossible.

When physical flow models are used to study other phenomena, other force ratios must be satisfied in addition to those already mentioned. For the study of inclusion particle movement, for example, it is important to match the force ratios involving inertia, drag and buoyancy. This generates several other conditions to satisfy, such as matching the terminal flotation velocity, which is:

$$v_T \equiv \frac{g(\rho - \rho_p) d_p^2}{18\mu(1 + 0.15 Re^{0.687})} \quad (3-7)$$

Where v_T is particle terminal velocity, m/s; ρ, ρ_p are liquid, particle densities, kg/m^3; d_p is particle diameter, m; μ is liquid viscosity, kg/(m·s); g is gravity accel, $g = 9.81$m/s^2; Re is particle Reynold's number, $Re = \rho v_T d_p/\mu$.

In a full-scale water model, for example, 2.5mm plastic beads with a density of 998kg/m^3 might be used to simulate 100μm 2300 kg/m^3 solid spherical inclusions in steel because they have the same terminal flotation velocity(equation (3-7)), but are easier to visualize.

Sometimes, it is not possible to match all of the important criteria simultaneously. For example, in studying two-phase flow, such as gas injection into liquid steel, new phenomena become important. The fluid density depends on the local gas fraction, so flow similitude requires additional matching of the gas fraction and its distribution. The gas fraction used in the water model must be increased in order to account for the roughly fivefold gas expansion that occurs when cold gas is injected into hot steel. Adjustments must also be made for the local pressure, which also affects this expansion. In addition to matching the gas fraction, the bubble size should be the same, so force ratios involving surface tension, such as the Weber number, should also be matched. In attempting to achieve this, it may be necessary to deviate from geometric similitude at the injection point and to wax the model surfaces to modify the contact angles, in order to control the initial bubble size. If gas momentum is important, such as for high gas injection rates, then the ratio of the gas and liquid densities must also be the same. For this, helium in water is a reasonable match for argon in steel. In many cases, it is extremely difficult to simultaneously match all of the important force ratios. To the extent that this can be approximately achieved, water modeling can reveal accurate insights into the real process.

3.4.2 Computational models

In recent years, decreasing computational costs and the increasing power of commercial modeling packages are making it easier to apply mathematical models as an additional tool to understand complex materials processes such as the continuous casting of steel. Computational models have the advantage of easy extension to other phenomena such as heat transfer, particle motion and two-phase flow, which is difficult with isothermal water models. They are also capable of more faithful representation of the flow conditions experienced by the steel. For example, there is no need for the physical bottom that interferes with the flow exiting a strand water model, and the presence of the moving solidifying shell can be taken into account.

Models can now simulate most of the phenomena important to continuous casting, which include: The slag allows more uniform heat transfer to the copper wall. fully-turbulent, transient fluid motion in a complex geometry (inlet nozzle and strand liquid pool), affected by argon gas bubbles, thermal and solutal buoyancies, thermodynamic reactions within and between the powder and steel phases, flow and heat transport within the liquid and solid flux layers, which float on the top surface of the steel, dynamic motion of the free liquid surfaces and interfaces, including the effects of surface tension, oscillation and gravity-induced waves, and flow in several phases, transport of superheat through the turbulent molten steel, transport of solute, transport of complex-geometry inclusions through the liquid, including the effects of buoyancy, turbulent interactions, and possible entrapment of the inclusions on nozzle walls, gas bubbles, solidifying steel walls, and the top surface, thermal, fluid and mechanical interactions in the meniscus region between the solidifying meniscus, solid slag rim, infiltrating molten flux, liquid steel, powder layers and inclusion particles, heat transport through the solidifying steel shell, the interface between shell and mould (which contains powder layers and growing air gaps), and the copper mould, mass transport of powder down the gap between shell and mould, distortion and wear of the mould walls and support rolls, nucleation of solid crystals, both in the melt and against mould walls, solidification of the steel shell, including the growth of dendrites, grains and microstructures, phase transformations, precipitate formation, and micro segregation, shrinkage of the solidifying steel shell due to thermal contraction, phase transformations and internal stresses, stress generation within the solidifying steel shell due to external forces (mould friction, bulging between the support rolls, withdrawal, gravity), thermal strains, creep, and plasticity (which varies with temperature, grade and cooling rate), crack formation, coupled segregation, on both microscopic and macroscopic scales.

The staggering complexity of this process makes it impossible to model all of these phenomena together at once. Thus, it is necessary to make reasonable assumptions and to uncouple or neglect the less-important phenomena. Quantitative modeling requires incorporation of all of the phenomena that affect the specific issue of interest, so every model needs a specific purpose. Once the governing equations have been chosen, they are generally discretized and solved using finite difference or finite-element methods. It is important that adequate numerical validation be conducted. Numerical errors commonly arise from too coarse a computational domain or incomplete convergence when solving the nonlinear equations. Solving a known test problem and conducting mesh refinement studies to achieve grid independent solutions are important ways to help validate the model. Finally, a model must be checked against experimental measurements on both the laboratory and plant scales before it can be trusted to make quantitative predictions of the real process for a parametric study.

3.4.3 Fluid flow models

Mathematical models of fluid flow can be applied to many different aspects of the continuous casting process, including ladles, tundishes, nozzles and moulds. A typical model solves the following continuity equation and Navier-Stokes equations for incompressible Newtonian fluids, which are based on conserving mass (one equation) and momentum (three equations) at every point in a computational domain:

$$\frac{\partial v_i}{\partial x_i} = 0 \tag{3-8}$$

$$\frac{\partial}{\partial t}\rho v_j + \frac{\partial}{\partial x_i}\rho v_i v_j = -\frac{\partial P}{\partial x_i} + \frac{\partial}{\partial x_i}\mu_{\text{eff}}\left(\frac{\partial v_i}{\partial x_i} + \frac{\partial v_j}{\partial x_j}\right) + \alpha(T_0 - T)\rho g_j + F_j \tag{3-9}$$

Where $\partial/\partial t$ is differentiation with respect to time, s^{-1}; ρ is density, kg/m^3; v_i is velocity component in x_i direction, m/s; x_i is coordinate direction, x, y, or z, m; P is pressure field, N/m^2; μ_{eff} is effective viscosity, $kg/(m \cdot s)$; T is temperature field, K; T_0 is initial temperature, K; α is thermal expansion coefficient, $m/(m \cdot K)$; g_j is magnitude of gravity in j direction, m/s^2; F_j is other body forces, e.g., from electromagnetic forces; i,j is coordinate direction indices.

The second-to-last term in equation (3-9) accounts for the effect of thermal convection on the flow. The last term accounts for other body forces, such as due to the application of electromagnetic fields. The solution of these equations yields the pressure and velocity components at every point in the domain, which generally should be three-dimensional. At the high flow rates involved in these processes, these models must incorporate turbulent fluid flow. The simplest yet most computationally demanding way to do this is to use a fine enough grid (mesh) to capture all of the turbulent eddies and their motion with time. This method, known as "direct numerical simulation," was used to produce the instantaneous velocity field in the mould cavity of a continuous steel slab caster. The 30 seconds of flow simulated to achieve these results on a 1.5 million-node mesh required 30 days of computation on an SGI Origin 2000 supercomputer. The calculations are compared with particle image velocimetry measurements of the flow in a water model. These calculations reveal structures in the flow pattern that are important to transient events such as the intermittent capture of inclusion particles.

Flow in the mould is of great interest because it influences many important phenomena that have far-reaching consequences on strand quality. Some of these phenomena are illustrated in Fig. 3-12. They include the dissipation of superheat by the liquid jet impinging upon the solidifying shell(and temperature at the meniscus), the flow and entrainment of the top-surface powder layers, top surface contour and level fluctuations, and the entrapment of subsurface inclusions and gas bubbles. Design compromises are needed to simultaneously satisfy the contradictory requirements for avoiding each of these defect mechanisms.

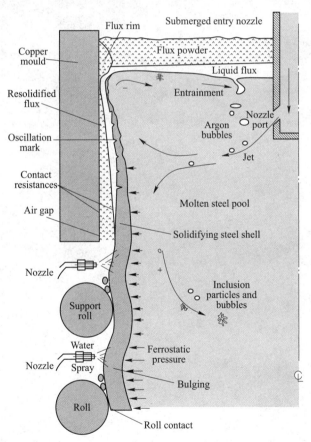

Fig. 3-12 Schematic of phenomena in the mould region of a steel slab caster

It is important to extend the simulation as far upstream as necessary to provide adequate inlet boundary conditions for the domain of interest. For example, flow calculations in the mould should be preceded by calculations of flow through the submerged entry nozzle. Nozzle geometry greatly affects the flow in the mould and is easy to change, so it is an important subject for modeling.

The flow pattern changes radically with increasing argon injection rate, which requires the solution of additional equations for the gas phase, and knowledge of the bubble size. The flow pattern and mixing can also be altered by the application of electromagnetic forces, which can either brake or stir the liquid. This can be modeled by solving the Maxwell, Ohm and charge conservation equations for electromagnetic forces simultaneously with the flow model equations. The great complexity that these phenomena add to the coupled model equations makes these calculations uncertain and a subject of ongoing research.

4 Continuous Casting Process

4.1 Foreword

After controlling the composition and temperature, and removing nonmetallic inclusions, the molten steel(heat) is transferred at a ladle support turret and pours into tundish and diverted at any point up to the caster, where it solidifies to produce semi-finished or finished products.

The process is started by plugging the bottom of the mould with dummy bar. Once in the mould, the molten steel freezes against the water-cooled walls of a bottomless copper mould to form a chilled shell. The mould is oscillated vertically in order to discourage sticking of the shell to the mould walls. Drive rolls lower in the machine continuously withdraw the shell from the mould at a rate or casting speed that matches the flow of incoming metal, so the process ideally runs in steady state.

The most critical part of the process is the initial solidification at the meniscus, found at the junction where the top of the shell meets the mould and the liquid surface. This is where the surface of the final product is created, and defects such as surface cracks and inclusion traps can form, if problems such as level fluctuations occur.

Below mould exit, the thin solidified shell(6-20mm thick) acts as a container to support the remaining liquid, which makes up the interior of the strand. Water or air mist sprays cool the surface of the strand between the support rolls. After the center is completely solid the strand is cut with oxyacetylene torches into products of any desired length.

After the steel semis leaves the caster with certain length, it is reheated to a uniform temperature and rolled into the selection of standard structural shapes of sheet, bar, rail, and other shape.

4.2 Treatment of Liquid Steel

For productivity and quality reasons, there is a trend in modern steelmaking to transfer time-consuming operations like temperature adjustment, deoxidation and alloying from the furnace to ladle treatment stations. These treatments are particularly important where the continuous casting process is involved because temperature and composition must be closely controlled.

4.2.1 The temperature control

The temperature control of the molten steel as it enters the mould needs to be more accurate in the continuous casting process than in conventional casting. Too high a superheat may lead to breakouts and produces a fully dendritic structure which is often associated with poor internal quality. On the other hand, too low a temperature may cause casting difficulties due to nozzle clogging and result in dirty steel.

The steel temperature in the tundish normally lies between 5℃ and 20℃ above the liquidus for slab casting and between 5℃ and 50℃ for billet or bloom casting; this differential depends on steel grade and, for example, is about 45℃ for stainless steel slab casting from small furnace.

In order to keep the steel temperature within the prescribed limits during the whole cast, temperature uniformity in the ladle is of paramount importance. Stirring is required before casting in order to destroy the temperature stratification in the ladle. Rinsing has become a common practice; the steel is flushed with either nitrogen or argon, injected by means of a porous plug at the bottom of the ladle or through a hollow stopper rod at a separate rinsing station.

Fig. 4-1 illustrates how to bring the temperature of the majority of heats to within +7℃ to 0℃ of a defined ladle target for line pipe X70 steels.

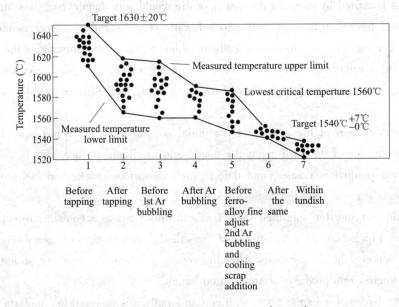

Fig. 4-1 Example of temperature control procedure

4.2.2 The composition control

Control of chemical composition can be performed during vacuum or rinsing treatments. On the basis of the analysis of a sample or of an electromotive force oxygen activity measurement made after homogeneity of the metal is attained, trimming additions can be calculated to ensure correct deoxidation. The best way to introduce trim deoxidants is at a high velocity (powder injection with inert bath gas, wire feeding or bullet shooting) while stirring the bath. Decreasing the need for alloys by careful exclusion of furnace slag from the ladle simplifies trimming.

Vacuum treatment is versatile but its use is not mandatory for good ladle metallurgy. Low pressure treatment, however, is the only way to remove hydrogen before casting or to decarburize to extremely low levels.

Desulphurization can also be carried out during ladle treatment after deoxidation. Use is made of calcium and magnesium base compounds. These elements not only have a fair desulphurizing power when added under good conditions (deep injection in killed steel) but they also transform the remaining sulphur into non-plastic inclusions and they can favor oxide elimination. Another advantage of calcium treatment is the transformation of alumina inclusions into liquid calcium aluminates which leads to reduced nozzle clogging.

To retain the advantages of ladle metallurgy it is important to keep in mind the deleterious effects of subsequent metal re-oxidation. Particular attention must be paid to the reduction of furnace slag entrainment, the nature of the ladle lining material, the last grade made in the ladle and the decrease of air entrainment in the casting streams.

4.2.3 Impurities control

Inclusions can generate many defects in the steel product. For example, Low Carbon Al-Killed steel cans suffer from cracked flanges due to lack of formability, while axels and bearings suffer fatigue life problems. Both formability and fatigue life are highly affected by sulfide and oxide inclusions in the steel. Sliver defects occur as lines along the steel strip surface parallel to the rolling direction. Slivers plague low carbon Al-killed steel sheet for automotive applications, causing both cosmetic surface imperfections and formability problems. They usually consist of aluminates originating from deoxidation or from complex non-metallic inclusions from entrained mould slag.

Tap oxygen content is measured during tapping the ladle or before deoxidant addition. The tap oxygen content is typically high, ranging from 250ppm to 1200ppm for IF production.

Aluminum additions to deoxidize the melt create large amounts of Al_2O_3 inclusions. This suggests that a limitation on tap oxygen content should be imposed for clean steel grades. However, as shown in Fig. 4-2, there is no correlation between tap oxygen and steel cleanliness. This is consistent with claims that 85% of the alumina clusters formed after large aluminum additions readily float out to the ladle slag, and that the remaining clusters are smaller than 30μm.

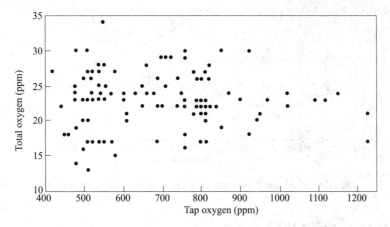

Fig. 4-2   Tap dissolved oxygen and final total oxygen in ladle

Naturally, the decision to ignore tap oxygen depends on the time available to float inclusions and on the availability of ladle refining, which can remove most of the generated inclusions. Fig. 4-3 shows how RH treatments reach the same final total oxygen level, regardless of initial tap oxygen, so long as the degassing time is long enough, "say", 15min. A final consideration is that the tap oxygen content affects the decarburization rate, especially for producing ultra low carbon steel.

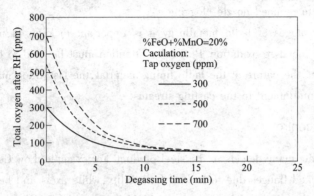

Fig. 4-3 Effect of tap oxygen on total oxygen removal in ladle during RH degassing

Calcium treatment of low carbon Al-killed steel is attractive because it can liquefy the oxides and sulphides in molten steel and modify their shape and deformability in the solidified steel. The liquid calcium aluminates coalesce and rise more easily than the clusters of solid alumina inclusions. This facilitates their removal to the slag and lowers total oxygen, in addition to avoiding nozzle clogging. To achieve liquid inclusions, the calcium must be present in the correct proportion. The acceptable range is narrow and depends on the alumina content, as documented by the equilibrium phase diagram. In addition, the sulfur content must be low in order to maintain liquid inclusions over the range of Al contents found in Al-killed steel, as shown in Fig. 4-4. Because Ca is so reactive, it is only effective after the steel has been deoxidized and if slag entrapment can be avoided.

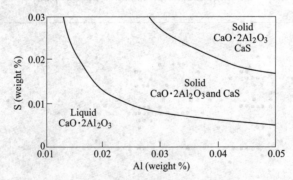

Fig. 4-4 The effect of Al and S contents on inclusions in equilibrium with Fe-Al-S melts

4.3 Tundish to Mould

Molten steel is poured from a casting ladle via a tundish into an open water-cooled copper mould. At first the bottom of the mould is closed off by a staring-bar, which then leads transport of the hot strand from the mould into the continuous withdrawing rolls. The strand, which starts to solidify in the mould, passes through a cooling system before it finally reaches the withdrawing rolls. The starting-bar is separated from the hot strand before or after it reaches the parting device. The latter, which may either be a flame-cutter or hot shears, moves at the same rate as the hot strand and cuts it into the lengths required.

The purpose of the tundish is to feed a defined quantity of molten steel into one or more moulds. This can be done by using nozzles controlled by stopper, slide-gates, or other means. The tundish may initially be cold, warm, or hot according to the nature of its refractory lining. The mould not only forms the strand section but also extracts a defined quantity of heat, so that the strand shell is strong enough for transport by the time it reaches the mould-outlet. The mould may be made from copper tube or hardenable copper alloy, depending on the shape and size of the strand to be cast. As a rule, tubular moulds are used for smaller sections. The interior surface of the mould may be coated with chromium or molybdenum to reduce wear and to suit heat-transfer from the alloy being cast. The mould is tapered to match steel-shrinkage and the type of steel concerned. Moulds used today range from 400mm to 1200mm in length overall, but their usual length is between 700mm and 800mm. The problem of steel adhering to the mould-sides is usually countered by oscillating the mould sinusoidally relative to the strand and by adding lubricant(oil or casting flux) in an attempt to cut friction between the mould and the steel. The lubricant, particularly casting-flux, has an additional metallurgical function. The choice of lubricant depends on the qualities required and the casting condition; it is particularly important that casting-flux should be chosen to match the quality-programme precisely.

The level of steel in the mould may be controlled manually or by an automatic system. Either method may be used to keep the level constant or to match the incoming molten steel, i. e. to accommodate variations in casting rate. Manual control is effected via the stopper in the tundish or by varying the output rate. An automatic control system may meter radioactivity or infrared radiation or measure temperature via a probe in the mould wall to determine the steel-level and compensate any changes by actuating the stopper-mechanism(for constant pouring rate) or controlling the withdrawing rolls(varying casting rate).

The type of starting-bar used for continuous-casting depends on the type of installation. Rigid starting-bars can be used in vertical system, while articulated dummy bars or flexible strip have to be used in bowed installations. The starting bar can be connected to be hot strand in different ways, one is by welding the fluid steel using a jointing element(flat slab, screw, or fragment of rail) which is soluble in the starting-bar; another is by casting the connector in a specially shaped head in the dummy bar in a way that enables it to be released by unlatching.

The thickness of the solidified strand shell on leaving the mould depends first of all on how long the steel is in contact with the mould, but it also depends on the specific thermal conductivity of the mould and on the amount of superheat that steel has when it enters the mould. It can be determined with fair accuracy using the following parabolic formula(4-1):

$$C = K\sqrt{t} \tag{4-1}$$

Where C is the thickness of the strand shell (mm); K is the solidification characteristic (mm/$\min^{1/2}$); t is the solidification time (min).

The solidification characteristic in and near the mould lies between 20 and 26, depends on the operating conditions. The thickness of the solidified strand shell on leaving the mould is about 8%-10% of the strand-thickness, depending on casting rate. A secondary cooling-area under the mould speeds up completion of the solidification process. The coolant usually is water but a water/air mixture or compressed air is also sometimes used. The secondary cooling area is divided into several zones to suit coolant flow rates. The necessary quantity of water is sprayed over the entire strand by spray-bars. The ferrostatic pressure may be so high in relation to the strand cross-section and the casting rate that the strand has to be supported to prevent bucking. The equipment for this is expensive in plants producing blooms and especially slabs.

In open stream casting the liquid metal flows directly, through the air, from the ladle to the tundish or from the tundish to the mould. Under these conditions the unprotected metal stream picks up oxygen (and some nitrogen) from the air and deleterious inclusions are formed in the liquid steel. These inclusions are transferred into the casting mould where they are either retained within the cast section or float to the surface of the liquid steel. Those present on the liquid steel surface are subsequently trapped in, the solidifying shell and either result in surface defects on the product in rolling or a catastrophic break in the shell below the mould. In addition to the direct formation of inclusions in the exposed steel stream, air entrained in the stream can also react with liquid steel both in the mould and tundish.

To avoid these problems shrouded stream casting is employed. Emphasis was first placed on shrouding the metal stream between the tundish and mould because of severity of the problem. However, ladle to tundish stream shrouding is now widely employed, especially in slab casting of aluminum killed steels where the prevention of alumina inclusions is of paramount importance. There are two basic types of shrouding with numerous variations and combinations:

(1) Gas shrouding;
(2) Refractory tube shrouding.

Gas shrouding is frequently used in casting small sections on billet machines because of operating difficulties experienced with refractory tubes: there is insufficient space to introduce a tube without encountering metal freezing between the mould wall and tube.

There is a variety of designs including: the Pollard steel tube shroud in which gas is introduced at the mid-point of the tube at low velocity and exits between the tube and nozzle, and between the tube and mould; complete enclosure between the tundish and mould using a flexible coupling; truncated pyramidal enclosures; and a liquid nitrogen curtain (Fig. 4-5). Nitrogen or argon is used as the protection gas. Gas shrouding alone is not commonly used for preventing oxidation of the ladle to tundish stream. However, one design in use employs a circular ring which is attached to the ladle at one end and is sealed by a sand seal at the other end when the ladle is lowered toward the tundish thus forming an enclosed box: the box is then pressurized with argon.

Refractory tube shrouds are commonly used for casting aluminum killed steel. They are used

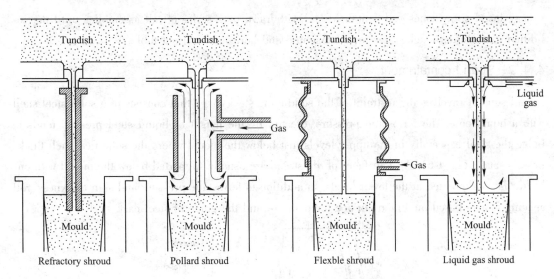

Fig. 4-5 Configurations for shrouding from ladle to mould

both between the ladle and tundish, and tundish and mould. One end of the tube is attached to the ladle (or tundish) with the other end immersed in the steel when the tundish for mould is filled with metal. Refractory tubes are usually made of fused silica or alumina graphite.

The mechanical design of the refractory tube is important, especially at the exit end that is immersed in the steel. One type is a straight through design. Another type, generally used in the mould, has a multi-port (opening) design, such as a bifurcated tube with the bottom of the tube closed and two side openings located near the bottom of the tube. This type of shroud avoids deep penetration of the pouring stream into the creator of the solidifying strand and modifies the flow pattern in the mould. Thus, the inclusions in the pouring stream are not entrapped in the solidifying section but rise to the surface of the liquid metal and are removed with the slag formed by the mould powder.

In many plants, the design of the shroud attachment includes the capability for replacing a worn shroud so that along sequence of heats can be cast without interruption.

At some plants, argon is introduced into the refractory tube to avoid aspiration of air through pores and joints that is caused by the venturi effect of a moving metal stream.

4.4 Secondary Cooling

4.4.1 The control of secondary cooling

In modern slab casting machines, secondary cooling, strand containment and withdrawal form a closely integrated and interlocked system that also includes strand bending and straightening. In the older designs of billet and bloom casting machines, there was a greater functional as well as physical separation of the components of this part of the casting operation.

The system is designed to produce a final cast section that has the proper shape, and internal and surface quality. To accomplish these results the solidifying section leaving the mould is cooled in a series of spray zones and contained and withdrawn by a series of roll assemblies until the solidified cast section reaches the cut off machine and horizontal run out table.

4.4.2　Strand containment

Strand support involves the restraint of the solidifying steel shape that consists of a solid steel shell with a liquid core. The ferrostatic pressure, created by the height of liquid steel present, tends to bulge the steel especially in the upper levels just below the mould where the solidified shell thickness is small(Fig. 4-6). All four faces of the strand are usually supported below the mould with only two faces supported at the lower levels. In addition to ferrostatic pressure and skin thickness, roll spacing is also based on strand surface temperature and the grade of steel cast.

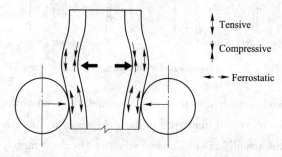

Fig. 4-6　Stresses in the solidifying skin due to ferrostatic pressure

　　Strand bending and straightening in addition to contain the strand, the series of rolls that guide the strand through a prescribed arc from the vertical to the horizontal plane must be strong enough to withstand the bending reaction forces. During bending, the outer radius of the solid shell is placed in tension and the inner radius in compression. The resulting strain, which is a function of the radius of the arc and the strength of the particular grade of steel being cast, can be critical; excessive strain in the outer radius will result in metal failure and surface defects(cracks). To minimize the occurrence of surface defects but, at the same time, maintain a minimum effective arc radius, triple point bending has recently been introduced(i. e. ,three arcs, with progressively smaller radii).

　　A multi roll straightener is installed following the completion of bending which, as the name implies, straightens the strand and completes the transition from the vertical to horizontal phase. During straightening the strand is unbent which reverses the tension and compression forces in the horizontal faces of the strand.

4.4.3　Strand withdrawal

The strand is drawn through the different parts of the casting machine by drive rolls. Drive roll system is designed to produce compression forces in the surface of the strand to enhance the surface quality. Thus, the objective is to push the strand through the casting machine, as opposed to pilling the strand with the attendant tensile stresses that tend to produce surface defects. In all cases, the pressure exerted by the drive rolls to grip the strand must not be excessive; excessive pressure will deform the shape of the section being cast.

Following straightening, the strand is conveyed on roller tables to a cut off machine where the section is cut to the desired length. The cast product is then either grouped or transported directly to the finishing mills or, in the case of billets, to cooling beds which are predominantly of the walking beam type to maintain product straightness.

4.4.4 Cooling water system

Control of heat transfer in the mould is accomplished by a forced convection cooling water system, which must be designed to accommodate the high heat transfer rates that result from the solidification process. In general, the cooling water enters at the mould bottom, passes vertically through a series of parallel water channels located between the outer mould wall and a steel containment jacket, and exits at the top of the mould.

The primary control parameters are: The volume of water at the required water temperature, pressure and quality. The flow velocity of water uniformly through the passages around the perimeter of the mould liner.

4.4.4.1 Water volume, temperature, pressure and quality

Typically, a pressurized recirculating closed loop system is employed. The rate of water flow should be sufficient to absorb the heat from the strand without an excessive increase in bulk water temperature. A large increase in temperature could result in a decrease in heat transfer effectiveness and higher mould temperatures. For this same reason, the inlet water temperature to the mould should also not be excessive; a proper mould water pressure is also required. For example, as discussed previously, higher water pressures tend to suppress boiling but excessively high pressures may cause mechanical mould deformation.

Water quality is an important factor with regard to scale deposition on the mold liner. Scale deposition can be a serious problem because it causes an additional thermal resistance at the mould cooling water interface that increases the mould wall temperature leading to adverse effects such as vapor generation and a reduction in strength of the copper liner. The type and amount of scale formed is mainly dependent on the temperature and velocity of the cooling water, the cold face temperature of the mould, and the type of water treatment.

4.4.4.2 Water flow velocity

To achieve the proper flow velocity, the cooling system is designed such that the velocity is high enough to produce an effective heat transfer coefficient at the mould cooling water interface. Too low a flow velocity will produce a higher thermal resistance at this interface, which may lead to boiling and its adverse effects. In general, the higher the cooling water velocity, the lower is the mould temperature. The cooling system should also be designed to maintain the required flow velocity distribution uniformly around the mould and to maximize the area of the faces that are directly water-cooled. Uniform flow distribution can be achieved by the proper geometrical design of the water passages with the use of headers and bale plates.

Monitoring the operating parameters of the mould cooling system provides an assessment of the casting process. For example, with a constant cooling water flow rate, the heat removed from a mould face will be directly related to the difference between the inlet and outlet water temperature, AT. Thus, an excessively large DT may indicate an abnormally low flow rate for one or more mould faces, whereas an excessively small DT may indicate an abnormally large scale buildup for one or more mould faces. An unequal DT for opposite faces may result from an unsymmetrical pouring stream mould distortion, or from strand misalignment.

4.4.5 Casting speed control

Together with sequence casting, casting speed and high machine availability determine output from a continuous casting plant. Large differences are found in the casting speeds recorded in the survey. The restrictive conditions such as steel grade, machine length and machine sophistication, differ for each caster.

Fig. 4-7 illustrates the broad scatter that exists in the relationship between slab thickness and casting speed.

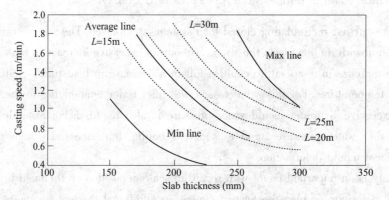

Fig. 4-7 Casting speed as a function of slab thickness

Such variations of performance are indicative of a rapidly developing technology. While there is a tendency for the more difficult analyses of steel to be cast at lower speeds, there are some exceptional performances on metallurgically demanding products and it is evident that future productivity from slab casters must be assumed to be higher than generally had been thought possible in the past.

In Fig. 4-8 and Fig. 4-9, the lines drawn in have been calculated as follows.

For bloom/billet casters (Fig. 4-8) the deviations are much smaller than for slab casters, the range from maximum to minimum casting speed being a factor of 2. As illustrated before, a notable exceptions relate to US Steel's plants.

The casting speed for rail blooms is limited to prevent break-outs and surface cracks. While the performance statistics for slab casters indicate a high potential for further development in casting

4.4 Secondary Cooling

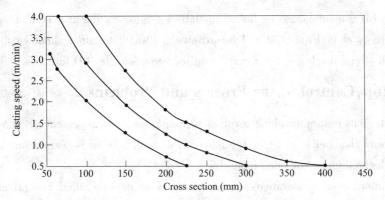

Fig. 4-8 Casting speed as a function of bloom/billet thickness

speed, billet and bloom casters, which already operate at very high speeds, may be approaching the limit where break-outs become too frequent to be tolerated and a new technology will have to be developed if casting speeds are to increase significantly.

Looking at the productivity (Fig. 4-9) in tons/minute/strand as a function of cross section of strand, the tonnage though slab casters naturally is much greater because of the larger cross-sectional area for a similar thickness. These data have been calculated in terms of the minimum, average and maximum speeds derived from the statistics provided.

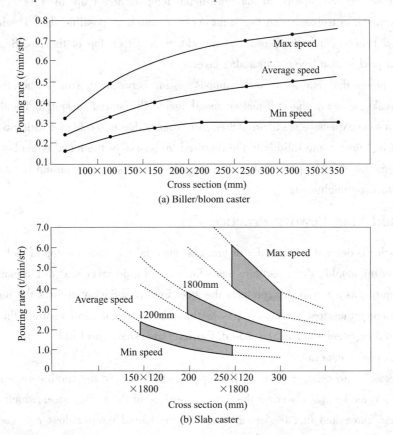

Fig. 4-9 Productivity of slab and billet/bloom casters as a function of thickness of strands

While large LD converters can be linked to slab casters(operating at maximum speed) in terms of the productivity of both units, it will be extremely difficult to run technological development; however 6 to 8-strand machines are operating on feed from smaller LD furnaces.

4.5 Startup, Control of the Process and Problems

Starting a continuous casting machine involves placing a dummy bar(essentially a metal beam) up through the spray chamber to close off the base of the mould. Metal is poured into the mould and withdrawn with the dummy bar once it solidifies.

Many continuous casting operations are now fully computer-controlled. Several electromagnetic and thermal sensors in the ladle shroud, tundish and mould sense the metal level, flow rate and temperature of the hot metal, and set the rate of strand withdrawal via speed control of the withdrawal rolls. Flow rate of hot metal through the shrouds is controlled by slide gates at the tops of the shrouds. The computer can also set the mould oscillation rate and the rate of mould powder feed, as well as the mould cooling rate(through control of the water flow).

While the large amount of automation helps produce castings with no shrinkage and little segregation, continuous casting is of no use if the metal is not clean beforehand, or becomes "dirty" during the casting process. One of the main methods through which hot metal may become dirty is by oxidation, which occurs rapidly at molten metal temperatures (up to 1700℃). To prevent oxidation, the metal is isolated from the atmosphere as much as possible. This is why shrouds are used for metal transfer, and why powder is introduced to sit on top of the mould pool. Often the tundish metal pool is also covered in a slag layer.

A major problem that may occur in continuous casting is breakout. This is when the thin shell of the strand breaks, allowing the still-molten metal inside the strand to spill out and foul the machine, requiring an expensive shutdown. Often, breakout is due to too high a withdrawal rate, as the shell has not had the time to solidify to the required thickness, or the metal is too hot, which means that final solidification takes place below the straightening rolls and the strand breaks due to stresses applied during straightening.

4.6 Mould Flux Powder Practice

A mould powder is defined as a powder or granular material added to the top of molten steel in the continuous casting mould, which partially melts forming a liquid layer next to the molten steel. This liquid, now known as a mould flux, protects the steel from reoxidation, absorbs the non-metallic inclusions, lubricates the steel shell as it passes through the mould, and controls the heat transfer from the solidifying steel shell to the mould. Particle size, shape, and bulk density will affect the flowability of powder material.

When the mould powder is applied on the top of molten steel in the continuous casting mould, it has to flow over and completely cover the exposed surface of the molten steel, which is particularly important when automated flux feeders are used. If the mould powder does not readily flow then some steel surface will remain exposed. The result is insufficient thermal insulation and an increased reoxidation of the steel and a reduced tendency for the mould powder to absorb non-metallic inclusions.

4.6.1 Properties of mould powder

4.6.1.1 Mould powder selection

The mould powder is mostly a synthetic slag which is added on top of the strand during the casting process. Since the steel is liquid, the melting powder is drawn into the gap between strand the mould through the oscillatory movement of the mould. In order to satisfy quality requirements, the right powder has to be chosen, since the choice influences oscillation mark depth and mould powder consumption(influencing the strand lubrication). Several estimations for the mould powder consumption can be found in the literature, most of them taking the powder's viscosity and the mould oscillation settings into account. One estimation is:

$$Q_s = \frac{A}{\eta^{0.5}} \frac{1}{v_c} \cdot t_N + B \tag{4-2}$$

Where Q_s denotes the mould powder consumption; η is the powder's viscosity in Pa·s; with A and B as fitting parameters, depending on the exact plant layout.

4.6.1.2 Types of mould powder

Mould fluxes can be classified into various types. The most common types are:

(1) Fly ash powders-Mechanical blends in which powdered fly ash is a significant component of the mix. The availability of consistent fly ash has limited the production of this type of powder.

(2) Synthetic powders-Mechanical blends of fine powdered raw materials made with high shear mixing.

(3) Perfused or fritted fluxes-These fluxes have a sizeable portion that is pre-melted and sized.

(4) Granular fluxes-Spherical or extruded granules are designed to have less dust than powdered fluxes. Spherical granules are particularly suitable for automatic flux applications.

Mould fluxes are designed for specific steel grades and steel plant conditions. Actual mould flux chemistries vary greatly depending on properties required. Typical ranges are shown in Table 4-1.

Table 4-1 Typical chemistry ranges for mould fluxes

CaO	SiO_2	Al_2O_3	TiO_2	C	Na_2O	K_2O
25%-45%	20%-50%	0-10%	0-5%	1%-25%	1%-20%	0-5%
FeO	MgO	MnO	BaO	Li_2O	B_2O_3	F
0-6%	0-10%	0-10%	0-10%	0-4%	0-10%	4%-10%

4.6.1.3 Major functions of a mould flux

(1) Thermal insulation.

The flux must provide thermal insulation to prevent bridging, and steel floaters. Improved thermal insulation increases the temperature in the meniscus region of the steel which helps to make oscillation marks less severe and can reduce sub-surface defects such as pinholes. The main control of insulation is the density of the unreacted flux, but the type of carbon used and the physical condition of the flux can also affect the insulating properties. Too low a bulk density may cause an unfavorable dust problem. This problem is solved by a granulated mould flux. Due to its particle shape, spherical granular fluxes have improved flowability over other types of fluxes such as powders and extruded granules.

(2) Chemical insulation.

A continuous slag layer is effective in preventing oxidation of the steel by insulating the steel from the atmosphere provided the flux is low in reducible oxides.

(3) Inclusion absorption.

The liquid slag acts to absorb non-metallic inclusions such as alumina floating up out the steel. Fig. 4-10 shows how the basicity (CaO/SiO_2) of a flux affects the ability of a flux to absorb alumina. Alumina rods were immersed into the liquid fluxes with varying basicity ratios. The difference in the rod diameter (Delta R) was plotted against time for different fluxes. Normal flux basicity ratios range between 0.8 and 1.25 Absorption of non-metallics improves with increasing basicity and decreasing Al_2O_3 in the slag. A lower viscosity increases the kinetics of inclusion capture and dissolution. It should be noted that this ability of a flux is a compromise between the disire to produce clean steel and submerged entry nozzle life.

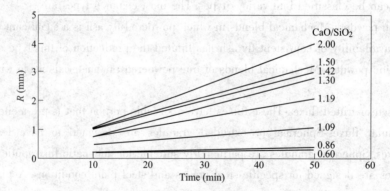

Fig. 4-10 Effect of CaO/SiO_2 on the rate of alumina dissolution

(4) Lubrication.

Lubrication is probably the most important function of a mould flux. The flux must act to provide a lubricating film between the solidifying shell and the water-cooled mould. A lower flux viscosity and/or solidification point tends to provide better lubrication, and this help prevent sticking.

An increase in basicity (CaO/SiO_2) as shown in Fig. 4-11, or an increase in the percentage of F or Na_2O will result in a greater tendency for the flux to recrystallize.

The crystallization index shown in Fig. 4-12 was established by obtaining a slag sample from the mould and measuring the opaque fraction of the cross sectional area of the solidified slag.

Index 0 indicates the slag was completely glassy while index 3 indicates the slag was 100% o-

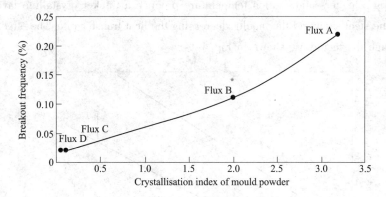

Fig. 4-11 Effect of crystallization index of mould powder on breakout frequency

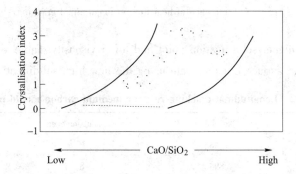

Fig. 4-12 Relation between basicity of mould flux on crystallisation index

paque. X-ray diffraction of the opaque area proved it to be crystalline. As the flux tends toward being more crystalline, the incidence of sticking can increase dramatically.

(5) Promotion of an even heat flow.

The final requirement is to provide an even heat flow. This is necessary to prevent uneven solidification of the steel shell which can lead to cracking of the cast product. Medium carbon steel grades have a larger shrinkage after solidification which makes these grades especially prone to cracking, as shown in Fig. 4-13.

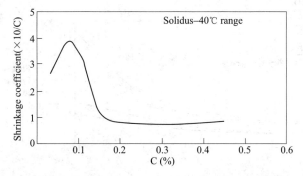

Fig. 4-13 Effect of carbon content on shrinkage after solidification

Fluxes having a higher solidification temperature produce a thicker crystalline layer in the flux film between the steel shell and the mould, decreasing the heat transfer rate. The effects that reduce heat flow through the mould are shown in Fig. 4-14.

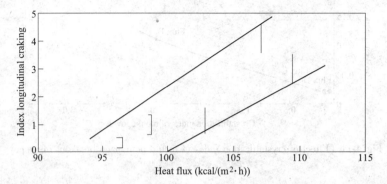

Fig. 4-14　Influence of heat flux on longitudinal cracking

The new flux has a higher solidification point and lower viscosity. Table 4-2 shows the improvement in the longitudinal surface cracking rate using the new high solidification point flux.

Table 4-2　Longitudinal cracking rate for medium carbon mould powders

Item	No. slabs	No. slabs cracked	Percentage
Old powder	1.485	75	5.05
New powder	894	8	0.89

4.6.1.4　Flux properties to consider

The viscosity, solidification point, melting point, and slagging speed are typically considered the most important properties of a flux. For some steel grades or conditions, it is sometimes necessary to give special considerations to the flux density and/or chemistry.

Viscosity is the major control of flux consumption with lower viscosities tending to cause an increase.

The effect of an increase in Al_2O_3 on the flux viscosity and solidification temperature is shown in Fig. 4-15 and Fig. 4-16.

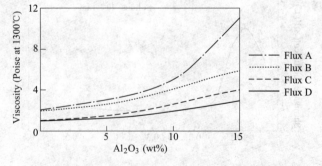

Fig. 4-15　Effect of increase in alumina on flux viscosity

4.6 Mould Flux Powder Practice

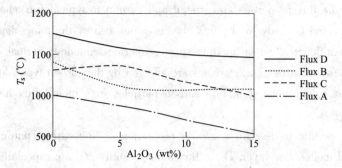

Fig. 4-16 Effect of increase in alumina on flux solidification temperature

Fluxes C & D show a much lower increase in viscosity and consequently, are more stable during usage. MgO additions to the flux are useful to stabilize viscosity as Al_2O_3 is absorbed.

The increase in Al_2O_3 in the molten flux chemistry is often below 3% with clean steel practices, such as slag deoxidation and good shrouding techniques. The combination of viscosity and solidification temperature effects the lubrication and heat transfer. A crystalline film is more porous than a glassy film which is a factor in the reduction in heat transfer. The type of mould plating, air gap and hydrogen levels also influence the rate of heat flow.

Fluxes are composed primarily of slag formers (or fillers), fluxing agents, and carbonaceous materials. The effect of an increase for various oxide additions on the viscosity and solidification point of fluxes are shown in Fig. 4-17 and Fig. 4-18.

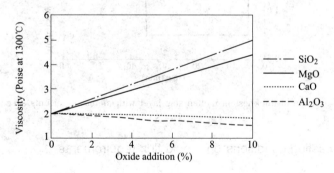

Fig. 4-17 Effect of oxide additions on flux viscosity

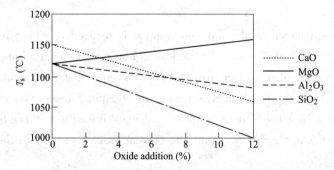

Fig. 4-18 Effect of oxide on flux solidification point T_s

The changes in the flux properties are general and pertain to typical flux chemistries. The flux melting point is affected not only by the flux chemistry but also by the mineralogical make up.

Carbon has a significant effect on melting speed, sintering tendency, thermal insulating properties, and slag rim. Flux melting characteristics are affected by the carbon type(due to difference in burning temperature). The amount of fine carbon particles in a mould flux helps determine its melting mode.

The alpha type melting mode with a greater proportion of fine carbon particles, form discrete droplets in the half molten slag layer. These fluxes melt rapidly and are especially suitable for high speed casting and unstable mould conditions. The beta type melting mode with a greater proportion of coarse carbon particles, form a partially sintered half molten slag layer. These fluxes supply slag slowly, steadily, and havebeen shown to be suitable for casting crack sensitive steels at low or intermediate casting speeds.

Fluxes(powder, fritted or granular) with an incorrect carbon system may have inadequate slag depths during speed changes. Fig. 4-19 shows a flux with the correct amount and types of carbon. An adequate liquid flux depth is maintained during and after the speed change.

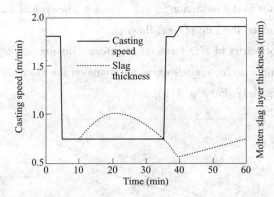

Fig. 4-19 Change in thickness of molten slag layer with increasing casting speed, Type A

4.6.2 Effect of casting conditions on mould flux requirements

The casting speed, oscillating cycle, and mould design are some of the key operating parameters that affect the design of a suitable mould flux.

(1) Casting speed.

As the casting speed increases, the flux consumption is reduced as shown in Fig. 4-20. If flux consumption is too low, a breakdown in the lubricating film may occur, resulting in sticking.

Low casting speeds may also have negative effects. The reduced steel flow rate can lead to cold spots in the mould which may result in steel floaters or bridging. Floaters may cause quality problems such as cracks or blow holes while complete bridging will lead to sticking. Special attention to the insulating properties of the flux is required to prevent problems at low casting speeds.

(2) Oscillation cycle.

High oscillation frequencies and shorter stroke lengths have reduced the oscillation mark depth

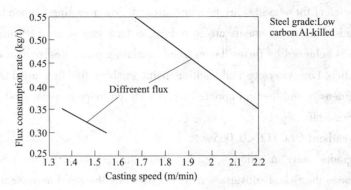

Fig. 4-20 Influence of casting speed on flux consumption

while helping to eliminate transverse cracking problems. A higher oscillation frequency reduces the negative strip time which in turn reduces the flux consumption in a similar way to higher casting speed. As a consequence, changes in oscillation cycles may require flux modification in order to prevent sticking.

(3) Mould design.

Mould plating has been adopted to improve slab surface quality by reducing star cracking. Mould life has also increased because of the wear resistant facing provided by plating. Chrome plating has often been replaced by nickel-based materials due to spalling problems. However, nickel has a significant negative effect on heat flow through the mould. A flux providing increased heat transfer may be needed.

4.6.3 Mould flux requirements and design by steel grade

(1) Low carbon aluminium killed (LCAK) (C<0.08%).

LCAK steels exhibit good high temperature mechanical properties, cracks not usually being a serious problem. However, high production rates require increased casting speeds. Requirements for these grades are good surface quality, good internal quality, and prevention of stickers. This is achieved by fluxes having good insulating properties, good absorption of non-metallics, good lubrication, and stable properties. Flux stability, the ability to absorb Al_2O_3 without an adverse effect on viscosity is very important, especially at higher casting speeds. A lower viscosity and/or solidification point helps the flux provide sufficient lubrication at higher casting speeds.

(2) Medium carbon (MED C) (C 0.08%-0.18%).

These grades are prone to cracking because of increased shrinkage associated with peritectic solidification. Prevention of longitudinal and transverse cracks are of paramount importance. This is achieved by fluxes that reduce heat flow through the mould and have a controlled melting speed.

High solidification point fluxes have been effective in reducing surface cracks. The high solidification point of the flux helps reduce heat flow through the mould while the low viscosity enables the flux to provide adequate lubrication.

(3) High carbon (HC) (C>0.18%).

The characteristics of these grades are poor hot strength, lower casting temperature, and typically cast at lower speeds. Flux requirements are to reduce surface slag scum, prevention of blow holes and stickers. This is achieved by fluxes having good insulating properties, correct carbon selection, and good lubrication. Low viscosity and melting point enables the flux to obtain rapid melting speed. A low flux density and the appropriate carbon addition helps achieve good insulation to prevent premature freeze off.

(4) Ultra low carbon(ULC)(C<0.005%).

These steel grades have a thinner mushy zone during solidification when compared to LCAK. This increases the risk of subsurface defects caused by the rapid movement of the solidification front. The addition of titanium to the steel can lead to a chemical reaction that alters the properties of the mould flux. The flux must be designed to provide good surface quality, good internal quality and prevent stickers. The requirements of the steel are achieved by fluxes having increased absorption of non-metallics, minimum carbon pickup, improved insulation, good lubrication, stable properties and minimal slag entrapment, and/or attachment.

High flux basicity along with special oxide compositions increase Al_2O_3 absorption and decrease slag entrapment and attachment. Carbon pickup can be minimised by maintaining sufficient liquid slag depth when combined with stable mould conditions. Alternatively, reducing the fixed carbon content of the mould flux can also be effective. Physical properties of the mould flux for each type of steel grade are summarized in Table 4-3.

Table 4-3 Mould flux properties by steel grade

Steel grade	Casting speed(m/min)	Viscosity(Poise)(1300℃)	T_s(℃)	T_2(℃)
LACK	0.8-1.4	2.5-1.0	1120-1060	1140-1060
	1.5-2.0	1.2-0.6	1060-960	1060-960
MEDC	0.8-1.2	5.0-0.8	1200-1100	1190-1120
	1.3-1.6	2.0-0.8	1180-1120	1180-1120
HC	1.3(max)	2.0-0.8		1180-980
ULC	0.8-1.2	2.6-1.2	1080-980	1120-1000
	1.3-1.8	1.5-1.0	1050-950	1040-1000

Marks: T_s—Solidification temperature; T_2—Melting temperature.

5 Defect and Quality Control of Continuous Casting Products

5.1 Foreword

Continuous casting process is not without of its problems. Considerable effort has been made by many researchers to establish adequate design, operation and maintenance of continuous casting machine to ensure metallurgical quality of the final product.

Good quality cast during continuous casting process can be produced when care is taken to avoid common casting defects. Some of these casting defects were observed during the casting process.

For example, a defect of rough and cracked cast surface occurs when the mould exit temperature becomes lower than the melting point of the melt. Subsequently, solidification takes place within the mould and results in rough cast surface which may lead to breakage.

Controversially, when the melt temperature is highly excessive, the solidification front moves far out of the mould exit, and may lead to metal breakout, and discontinue the process. Furthermore, when the solidification front passes into the cooling device, the cooler entrance may scratch the cast surface and form favored sites containing vapor bubbles which form when the cast surface touches the cooling water. Subsequently, a defect of periodic notches appears at the cast surface. These notches may be impacts on the cast surface of rapid collapse of the vapor bubbles, which end up at the bubbles' centers with a very high pressure, and with a very high inward velocity, i.e. above the speed of sound. This conforms with the audible periodic banging sounds which were heard during the occurrence of these periodic damaging notches to the cast surface. However, the periodic pattern of these notches may be due to periodic dry out and wetting waves within film or transition boiling regions at the cooler entrance.

Wavy surface defects, is initiated at the solidification front and results when the system is subjected to mechanical vibrations during the casting process.

The metal head inside the crucible influences the continuity of the process and the cast size, and must be equivalent to the losses during the melt traveling from the crucible to end up with no pressure at the mould exit. This metal head has to be properly harmonized with the casting speed and the ingot size so as to substitute for the metal consumed during the casting process. Otherwise, an excessive metal head will lead to melt breakout at the mould exit, or on the other hand, an insufficient metal head will lead to cast product with reduced or variable cross sections.

The cast quality is affected by the casting process variables. The key factor that determines the cast surface quality is the location of the solidification front, which has to be at or near the mould exit and is influenced by process variables such as mould exit temperature, cooling rate, mould-cooler distance, and casting speed. For example, increasing the casting speed results in locating and moving the solidification front out of the mould exit, and vice versa. This increase in the casting speed has to be compensated by an increase in the cooling rate and/or a decrease of mould-cooler distance. The accuracy of performing this; interferes with the continuity of the process and requires substantial level of personal skill and experience.

The defects met in continuously cast are inclusions, cracks, and shape defects.

(1) Inclusions vary with steel grade and deoxidation methods and rate affected by the equipment used by the caster operator and its design details. Control of inclusions is highly dependent on operational techniques.

(2) External and internal cracks are usually classified according to their geometrical appearance. Internal cracks may re-weld during rolling and pose fewer problems except with some grades. They are described as mid-radius and center-line cracks.

(3) Surface cracks are divided into longitudinal, mid-face and corner cracks, transverse mid-face and corner cracks and star cracks.

(4) Shape defects.

According to the different shape and size of casting products, it is divided into slab, bloom, billet and round bloom, etc. The shape defects of continuous casting products are not only the physical form problem, but also ties up to other problems as cracks and breakout.

5.2 Purity of Continuous Casting Products

5.2.1 The inclusions in the casting steel and its origin

5.2.1.1 Origins of inclusions

With the extension of application scope and the requirement of quality improving of continuous casts, more attentions have been paid to the characteristics and the damage of the inclusions. The inclusions have two prominent characteristics in continuous casting. First, its origin is complex. The molten steel has so many chances (tundish lining, stopper, nozzle, submerged entry nozzle, tundish and the mould powder in the tundish) to contact with the refractory from the tundish, a great area of steel exposed in air and oxidized seriously, which makes the cleanness of steel decreased; second, the floating up and separation of the inclusions in the tundish is very difficult.

Based on the practical conditions and technical process of A Special Steel Plant, several key techniques of production for the process of melting, refining and continuous casting have been developed by the industrial tests. And the following conclusions have been obtained in the study:

(1) The terminal carbon content must be controlled in the melting process in order to control the terminal oxygen content. And the ability of the dephosphorization for slag must also be increased.

(2) The slag with low basicity and low iron oxide (FeO) should be selected for the refining process. In order to adjust the aluminum content in the liquid steel in the range of 0.025%-4.04%, the value of aluminum as deposit deoxidizing oxygen should be increased. The argon

stirring with the flow rate of 180L/kg should be applied during heating in LF refining process.

5.2.1.2 The effect of different casting machine

With different continuous casting machines, the inclusions amount of the continuous casts is different observably. The inclusions amount with different casting machines of per kg casts is that:

Vertical: 0.04mg/kg.
Vertical with bending: 0.46mg/kg.
Curve type: 1.75mg/kg.
Horizontal continuous casting: 1.35mg/kg.

5.2.2 Measures to decrease the inclusions in the casting steel

As illustrated before, too low a temperature may cause casting difficulties due to nozzle clogging and result in dirty steel. Besides, calcium and magnesium base compounds used for desulphurization can also be carried out during ladle treatment after deoxidation. The advantage of calcium treatment is the transformation of alumina inclusions into liquid calcium aluminates which leads to reduced nozzle clogging.

Other helpful methods for improve the purity of semis is list as followings.

(1) Slag-free tapping for BOF.
(2) Refining technology.
(3) Non-oxidation casting technology.
(4) Tundish metallurgical purifier.
(5) High performance refractory.
(6) Electromagnetic stirring to protect device of teeming stream for ladle.

5.3 Surface Crack and Its Control

Common surface crack is shown in Fig. 5-1. It includes surface transverse corner cracks, longitudinal corner cracks, transverse surface cracks, longitudinal surface cracks, star cracks, oscillation marks, blowhole, and large inclusions, etc.

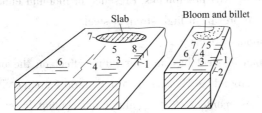

Fig. 5-1　Schematic diagram of surface cracks

1—Transverse corner cracks; 2—Longitudinal corner cracks; 3—Transverse surface cracks;
4—Longitudinal surface cracks; 5—Star cracks; 6—Oscillation marks; 7—Blow hole; 8—Large inclusions

5.3.1 Slab surface defects and controlling

Surface defects are a serious quality problem, which may require a slab conditioning or even scrapping.

(1) Longitudinal corner cracks.

Longitudinal corner cracks are seldom reported in slabs. Their formation seems to be due to inadequate support of the shell in the mould. Taper of mould, surface condition and mould water flow can be important. At US Steel's Texas Works the defect was eliminated when the 1,200mm long mould was replaced by a shorter 900mm mould.

(2) Mid-face longitudinal cracks.

It is generally accepted that mid-face longitudinal cracking of continuously cast slab is initiated in the mould and that the severity of cracking is influenced by the secondary cooling and slab support systems immediately below the mould.

(3) Star cracks.

Star cracks are generally associated with the presence of copper at the grain boundaries. They result from abnormal friction between mould and strand shell. Attention should be paid to mould surface quality and geometry (wear and deformation), lubrication (use a flux with higher fluidity), temperature of the mould wall (primary cooling efficiency) and contact time between mould strand shell (oscillation pattern).

Occurrence of this defect is enhanced by low casting speed (<0.7m/min). The best known method of avoiding star cracks is mould plating with chromium or explosion plating with nickel; however the latter is seldom used in slab casting.

(4) Transverse cracks and oscillation marks.

In some instance copper embrittlement produces transverse rather than star cracks.

In addition, at the unbending point, surface temperature is required to be outside of the low ductility through in order to avoid transverse surface cracking.

Transverse cracks may appear at the mid-face but they are most often corner cracks. They are usually associated with oscillation marks. These open when bending and straightening so the radius curvature of the strand affects crack frequency.

This type of cracking occurs in steel containing more than 0.020 percent aluminum and would be linked to embrittlement by AlN precipitates. Presence of niobium enhances the cracking sensitivity. This defect appears to prevail in high strength steel.

(5) Chemical composition.

The chemical composition of steel has an over-riding effect on the occurrence of longitudinal cracking: carbon content around 0.12%, low Mn/S ratio (around 20), high sulphur level (>0.025%), high manganese content (>1%), aluminum content lower than 0.004 percent and presence of niobium contribute to make a steel sensitive to cracking. The effect of the chemical composition of steel on cracking sensitivity may be due to extreme brittleness in the vicinity of the solidus. The critical temperature range is enlarged by the presence of alloying elements such as sulphur. The occurrence of the peritectic transformation in this critical range of temperature adds extra strain and increases the crack sensitivity of the solidified shell.

(6) Transverse surface cracks.

Transverse cracking in continuous casting is caused by longitudinal tensile strains. Such strains are generated in the spray chambers or by the straightening. The latter one appears to result in crack formation if the surface temperature is too low, particularly in the range of 700-900℃ corresponding to the low temperature zone of low ductility.

In order to evaluate the risk of transverse cracking, two simple indexes have been introduced. They are not explicit criteria in the sense that they do not explicitly predict crack initiation since no critical value is available(for example, yield stress could have been used to normalize indexes, if available). However, the value of indexes increases with the risk of transverse cracking so that it is possible to have a qualitative indication on the mechanical state of the slab in different situations. According to the objective of the study, the surface transverse cracks generally appear in the corners. The conclusions of the studies of Brimacombe and Mintz naturally lead to the simple and intuitive definition of indexes based on both longitudinal stress and longitudinal strain rate that have been identified as key factors in transverse crack formation. In addition, the two indexes combine also the metallurgical unfavorable aspects, i.e. the lack of ductility in the range of temperature between $T_1 = 700℃$ and $T_2 = 900℃$. They are established on the following bases:

1) Indexes are positives if mechanical state is favorable to open transverse cracks, equal to zero in other cases.

2) Indexes are significant when the material is characterized by low potential to deformation without cracking, i.e. when the ductility of the material is low.

3) The higher the indexes are, the higher the risk is.

The method to avoid the surface cracks in slab: Cracks initiation may be trace back to mould conditions and lubrication but crack propagation depends largely on secondary cooling conditions. To avoid cracking it is often advised to bend or straighten slab with a surface temperature above 900℃ or below 700℃. However, the cooling pattern along the strand has a prime effect on the occurrence of cracks; it has been suggested that cooling in the top spraying zone should be limited to less 100℃/m.

5.3.2 Surface defects in bloom and billet

Continuously cast blooms and billets routed to applications requiring a high surface quality have to be thoroughly inspected before further processing.

(1) Pinhole formation.

Pinhole formation results from gas evolution during early solidification. Hydrogen in the liquid steel must be controlled and can be lowered by vacuum degassing, to less than 3ppm. Nitrogen levels in the steel depend on the steelmaking practice. Carbon monoxide possibly plays the major role in blow hole. Control is a matter of achieving sufficient deoxidation taking into account the hydrogen and nitrogen contents.

(2) Blow hole.

The parameters which minimize blow hole formation are high casting temperature, high casting speed, submerged nozzle with dehydrated slag and high aluminium content.

Surface defects arising from inclusion are more frequent and more serious than those from pinholes and their suppression necessitation greater effort.

(3) Steel composition.

Occurrence of the surface defects depends on steel composition. The manganese-silicon steels give excellent results whereas oxide impurities are frequent in chrome-manganese steel with 0.2% carbon. As a rule, aluminium additions in excess of 0.020% significantly increase the probability of encountering inclusions unless oxygen is excluded.

(4) Longitudinal cracks.

Longitudinal cracks are very serious and may result in the scrapping of billets and blooms. The sensitivity of steel to surface cracking depends on its composition, sulphur content above 0.020% being especially deleterious.

The mould geometry (conicity, shape, surface quality) plays a leading part in surface crack formation.

A radical improvement of surface quality can be obtained with an immersed nozzle and a covering layer of slag in the mould. This technique can, in general, be applied to sizes greater 150mm. In some cases, it raises problems due to mould flux entrapment, resulting in sub-surfaces inclusions.

In the process of continuous casting, in order to produce the strand without surface defect, it is very important to control the initial steel solidification to deduce the corrugated surface vibrated mark. According to the theory of magneto hydrodynamics, the strand without surface defect can be produced through the "soft contact" between meniscus and the wall of crystalline cell using magnetic confinement of the alternative magnetic fields. In the process of the continuous casting, confirmation of the meniscus shape is not only the essential factor of the magnetic confinement computation, but also the key to achieving continuous casting successfully.

The process of slab continuous casting has been received considerable attention due to its shorter flow, lower investment, lower energy consumption and high productivity. Medium thin slab continuous casting machine is between traditional one and slab caster, and the quality of slab is relatively high. In the present work, the quality of medium slab produced by the machine in Anshan Iron and Steel Company was analyzed, and the properties of ladle top slag, LF refining process, wire feeding process, fluid flow in mould and continuous casting process for steel were investigated. The application of the optimization in actual process shows that the quality of slab has been improved obviously. The following conclusions can be draw:

(1) The basicity of top slag becomes higher after the addition of permeated flux with lime, and the content of (MnO+FeO) decreases, thus the sulphur recovery has been prevented that is in favor of LF refining.

(2) After the homogenization of temperature and chemical composition and desulphurization in ladle furnace, the quality of molten steel has been improved significantly that can meet the requirement for medium thin slab casting.

(3) The lifespan of mould copper plate has been improved obviously by increasing the flow rate of molten steel through the bottom hole of SEN and decreasing the strength of upward recirculation from side holes. The surface of molten steel in mould becomes more stable and the boiling of steel

flow near the narrow face becomes weak. The weight of molten steel cast one mould increases from 40,000t to 60,000t. In addition, the melting and flowing of mould flux have been improved significantly.

(4) The molten steel for medium thin slab casting becomes clean constantly by the deoxidation, LF refining, wire feeding, and tundish metallurgy, and the total amount of inclusions in slab can be reduced to 62ppm.

(5) In the production of steel, slab defects can decrease by reducing the casting speed, optimizing secondary cooling, using medium carbon flux with low heat conductivity and SEN with ϕ35mm ×38mm bottom hole. The occurrence of depression and transverse crack on wide face of slab decreases from two per meter to one per meter and the qualified slab has reached 98%.

5.4　Internal Defect and Its Controlling

Internal defect is shown in Fig. 5-2. It includes internal corner crack, centerline cracks, star crack and diagonal crack, loose, shrinkage, subcutaneous ghost line, and non-metallic inclusion, etc.

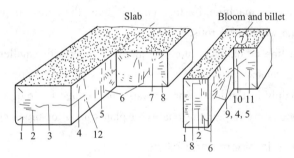

Fig. 5-2　Schematic diagram of surface cracks

1—Internal corner crack; 2—Intercolumnar crack; 3—Centerline cracks; 4—Centerline Macro-segregation;
5—Loose; 6—Intermediate crack; 7—Non-metallic inclusion; 8—Subcutaneous ghost line;
9—Shrinkage; 10—Star crack and diagonal crack; 11—Pinhole; 12—Gross segregation

5.4.1　Internal defects in slab

(1) Ghost lines.

The mid-radius cracks are described as radial streaks or ghost lines. These cracks are often filled with solute-rich liquid. They appear either as thin ribbon-like intergranular streak between the columnar grain structure of as-cast slabs when macro-etched, or as a close array of inter-connected streaks.

These cracks may be associated with mechanical deformations which induce tensile strains along the solidification front. One can distinguish between two main types of ghost lines:

1) Rows of parallel, thin streaks which extend perpendicularly to the narrow faces and which seem related to a lack of support of the stand.

2) Rows of parallel streaks or dispersed groups which grow perpendicularly to the wide face. These are produced by bending, straightening or rolling the strand shell on a liquid core, due

to bent, misaligned or seized rolls.

It is often stated that critical grades should not be straightened on a liquid core. Salzgitter AG experience would indicate, however, that such a practice has no adverse effect on internal quality provided it is not combined with additional tensile forces due to withdrawal.

The frequency of occurrence of radial streaks is closely related to the chemical composition of steel.

(2) Centerline cracks(Centerline segregation).

Centerline segregation or internal cracking can form if straightening or bending operations are carried out on a section with a liquid center.

Centerline cracking seems to result from bulging and/or roll misalignment which occurs near the point of the solidification. This defect may be traced back to the geometry and stability of the machine.

1) It may be due to a roll which is out of alignment in the vicinity of the end of the liquid pool.

2) It may appear when the liquid pool extends into the withdraw section as a result of some constructional instability of the machine. In this case it may then also depend on the set point hydraulic pressure of the withdrawal rolls.

Occurrence of centre line cracking is more frequent in slabs with smaller thickness. It also increases with the casting velocity. It does not appear to depend on casting temperature. Increasing the cooling intensity, mainly in the lower spray zones, may reduce its frequency.

Segregation may significantly impair the quality of plates used for pipes or welded structures.

5.4.2 Internal defects in blooms and billets

The internal quality of critical grades can be monitored by macro-etching or sulphur printing.

Internal cracks may occur as subcutaneous, mid-radial diagonal or center cracks. They are associated with the structure of the product and its correlated dendritic segregation. Frequently the cracks are filled with enriched liquid but, in some cases, they are open cracks(centre cracks).

The chemical composition of the steel has a strong influence on their occurrence.

The parameters which favor a high a high rate of internal cracking are: high casting temperature, high casting speed, excessive secondary cooling, poor alignment of support rolls and pinch rolls and excessive withdrawal pressure.

Adequate control of secondary cooling is essential when casting grades which are sensitive to cracking.

(1) Central porosity(Loose). Billets and blooms, mainly with square or round sections, often present central porosity. In practice two types of defects may be recognized.

(2) Shrinkage. "Cloud": these are numerous micro cavities dispersed in a central zone. This type of defect welds when the semis are rolled with a reduction ratio greater than 6 : 1 and it does not present any difficulties in the final use of the steel.

(3) Pinhole. "Hole": this is a single central cavity of variable diameter. Its formation results from solidification bridges and full dendritic structure. It disappears with reduction ratios around

10:1 if it has not been oxidized during reheating.

The incidence of this defect is favored by dendritic structure and high casting speed.

(4) Subsurface blow holes. So far, the only limitations appear to be free-cutting steel, leaded steel and 1%C-Cr for ball bearings. However, medium-carbon free cutting steel is normally cast as well as leaded steel. This latter grade shows an interior surface quality which requires 100 percent conditioning prior to further processing. It is often claimed that these grades can only be successfully cast by vertical machines.

One percent of C-Cr steel is believed to be one of the more difficult grades to be cast due to problems of internal cracking and carbide segregation.

5.4.3 The methods to decrease the internal defects

(1) Electromagnetic fields can be used to create strong, circulating flow patterns in metallic liquids. Theoretical methods of estimating flows in such systems have been developed.

In a continuously cast billet, a rotating field can initiate rotation of the metal in the liquid pool and in a continuously cast slab, liner fields are used to move the liquid metal along horizontal or vertical axes.

In billets casting, the detailed technology can differ but, generally, the electromagnetic stirrer is a water-cooled ring with induction coils surrounding the stand which are positioned at a level depending upon the billet size and the casting speed.

Electromagnetic stirring, which promotes an early transition from dendritic to equiaxed structure, is efficient in reducing or suppressing axial porosity independently of casting conditions.

Electromagnetic stirring to improve the as-cast structure of semis by the method developed by IRSID is used at a French steel plant where a four strand bloom caster is equipped with stirrers beneath the mould. Fewer defects on finished products and improved fatigue properties of automotive forgings have resulted.

In-line rolling decreases the central porosity before it can be oxidized during cutting to length. Electromagnetic stirring may compensate for a high casting temperature.

For slab casting, the application electromagnetic stirring is more recent.

The aim is the same, which is to say to break the dendrites along the solidification line and to decrease the width of the columnar zone. But due to the width of the slab, the effect is more difficult to achieve than it is for blooms or billets.

(2) Conditions favoring a columnar structure worsen segregation problem, i.e. high casting temperature and steel composition, in the ascending order 0.3%, 0.1% and 0.6% carbon. Larger sections and rectangular blooms display relatively less segregation.

As in slab casting, mechanical stresses applied in bloom and billet strands which are not completely solidified, can produce exaggerated segregation. Low levels of segregation can be achieved if the temperature of steel in the tundish is kept low.

(3) Internal cleanness is the most difficult problem to cope with in the continuous casting of

blooms and billets.

The improvements of cleanness in critical grades needs considerable development work. Equivalent cleanness levels to those found in conventional products can only be achieved by closely controlling deoxidation practice and casting conditions. In particular, for high grade application, i.e. shrouding the metal stream between ladle and tundish, submerged pouring under a slag cover in the mould and even a controlled atmosphere in the tundish.

Closed pouring is very difficult to achieve in billet casting(<150mm) where the only means of stream protection is a gaseous shroud. Moreover, the aluminium content in the liquid steel in the tundish must be low to prevent the small nozzle from the clogging and the excess aluminium needed has to be fed in the mould. This does not favor a high level of cleanness.

(4) Three factors, the solidification structure, the quality of the strand support and the length of liquid core have an overwhelming effect on segregation.

1) Heavy segregation has been shown to be associated with bulging of the slab. Casting critical grades for plates requires the tightest tolerances in the setting up of the machine; particularly roll alignment, spacing and deformation. The quality of the strand support depends on the roll pitch and diameter.

2) Absence of segregation is most often associated with the presence of a central equiaxed zone. In order to curtail segregation, the equiaxed region should be at least 30mm wide. The presence of this central equiaxed structure depends on chemical composition, superheat and, apparently, also thickness.

In the continuous casting process, superheat is the principal operating parameter which influences the final solidification structure. In most cases, a superheat of 15°C is aimed at in the tundish; a range of ±12°C is covered in 99% of cases. When casting critical grades(high strength steel) it is necessary to have better control of the tundish temperature in order to reduce the required superheat below 10°C.

An alternative means of producing a central equiaxed structure is electro-magnetic stirring.

1) In a general way, increasing casting speed favors stronger segregation. This is presumed to be the main reason why thinner slabs (160mm thick) consistently display heavier segregation than thicker sections. At lower casting speed the thicker solidified shell offers a better resistance to bulging at any given level in the casting machine.

2) An alternative means to strengthen the solidified shell and reduce segregation is to keep the solid shell colder by more intensive cooling in the bottom of the machine, extending to the final point of solidification.

In continuously cast billets and blooms, a central positive segregation is often associated with a fully dendritic structure. A particularly deleterious form of segregation is V-shaped lines which are thought to be formed by solidification bridges. This defect is undesirable for cold heading applications.

5.5 Shape Defects of Continuous Casting Products

The shape defects of continuous casting products are not only the physical form problem, but also ties up to other problems as cracks and breakout. Shape defects result either from uneven cooling of the face of the billets or from insufficient support of the strands.

5.5.1 Bulging deformation

5.5.1.1 Introduction

Bulging deformation are essential features for cast slabs, which influence the quality of flat products.

As shown in Fig. 5-3, the bulge forms on the billet shell under the combining action of the roller supporting force and the pressure of molten metal in the process of continuous casting, with the result that the solid-liquid inter-face cannot hold the planar interface. Kusano et al. found that the moving track of solid-liquid interface in the direction of casting is in the pattern of a sine wave formed by the periodic bending deformation of interface. This variation of interface has an influence on solute distribution in columnar crystal zone, and it also affects the formation of internal crack.

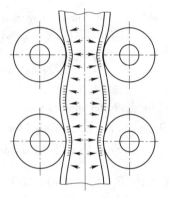

Fig. 5-3 Schematic diagram of bugle

When the solid-liquid interface occurs the periodic bending deformation, the change of the dendritic spacing in two-phase region is shown in Fig. 5-4. When the solid-liquid interface is planar, the dendrites in the two-phase region are shown in Fig. 5-4(a). When the solid-liquid interface occurs the bending deformation, the dendritic spacing will be changed in the continuous casting process. The spacing is maximal at the wave crest of the sine wave, as shown in Fig. 5-4(b). The spacing is minimal at the wave trough of the sine wave, as showed in Fig. 5-4(c). As mentioned above, the bending deformation of solid-liquid interface changes the temperature field, and the occurrence of temperature gradients along drawing direction provides the condition for the growth of secondary dendrites. Secondary dendrites will be produced when the primary dendritic spacing is large enough. The dendritic spacing increases obviously and the secondary dendrite grow quickly in the middle and final stage of solidification process. So it is possible that the growing secondary dendrites overlap when the interface reaches the wave trough of the sine wave, as shown in Fig. 5-4(d).

If the primary and the secondary dendrites are both developed, the closed area at the bottom of dendrites will be formed at the wave trough. If the enclosed region still exists at the wave crest and the molten metal of the enclosed region does not get filled during the solidification, the enclosed region will become a small porosity. If the porosity is not welded together in the subsequent deformation, it will become crack source.

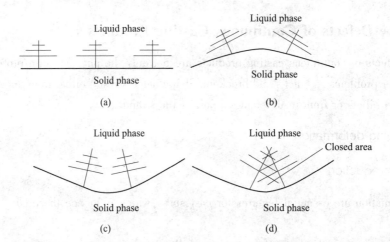

Fig. 5-4 Variation of dendritic spacing in the two-phase region with bending deformation of solid-liquid interface
(a) Solid-liquid interface is plana; (b) Solid-liquid interface locates in the wave crest of the sine wave;
(c) Solid-liquid interface locates in the wave trough of the sine wave;
(d) The closed area in formed at the wave trough of the sine wave

While the billet shell is relative thin, the primary dendritic spacing is very small and the secondary dendrites are not well developed. So the crack source is observed in the center, not on the surface of casting slab, as the dendrites do not overlap.

5.5.1.2 Slab's bulge size in continuous casting

The calculation formula of slab's bulges by utilizing the theory one plastic distortion of materials mechanics and creep deformation in continuous casting was obtained. The calculation formula and the effect of static pressure increment of liquid steel were considered. Thus, a new mathematic model was set up. The quantity calculations indicated that appropriated arrangement of second cooling water will reduce the size of bulges; The higher the velocity of pulling billet, the larger the size of bulges will be, the velocity of pulling billet varied in 0.1m/min, the corresponding size of bulges varied in the extent from 6% to 15% at the same position; The effect of the distance of rolls will be more obvious. The distance of adjacent rolls varied in 10mm, the corresponding size of bulges varied in the extent from 10.5% to 21% at the same position.

5.5.1.3 The methods to reduce bulge deformation

(1) Decrease the height of continuous casting machine.

(2) Low the rolling pitch in the secondary cooling zone.

(3) The disposal of rolling pitch should be from dense to sparse from the top of the casting machine.

(4) Backup roller should be centered assembled strictly and increase cooling intensity in the secondary cooling zone.

(5) Prevent the deformation of backup roller, the backup roller for slab should choose the multi-roller.

5.5.2 Rhomboidity of casting billet

5.5.2.1 Foreword

Rhomboidity of casting billet(off square) is also a defect in the continuous casting process. It could be called rhomboidity when the cross sectional diagonal is different as shown in Fig. 5-5.

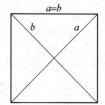
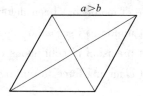

Fig. 5-5 Schematic diagram of rhomboidity

The quantities account of rhomboidity deformation would be expressed as $a-b(\mathrm{mm})$, or $R(\%)$.

$$R = \frac{a-b}{\frac{a+b}{2}} \times 100\% \quad \text{or} \quad R = \frac{a-b}{b} \times 100\%$$

When R is great to a certain degree, it would cause diagonal cracks of tow obtuse angle. The degree of R which caused diagonal cracks relates to hot ductility of steel. For example, low carbon steel(carbon content is 0.08%-0.12%), it would cause the diagonal cracks when $R>3\%$. It means that, low carbon billet would like to be rhomboidity but it will not cause the diagonal cracks when $R<3\%$. 1%C-1.5%Cr billet has the diagonal cracks even the rhomboidity is very low, the diagonal cracks and the rhomboidity appear in the same time. Ferritic stainless steel is easier cracking than austenitic stainless steel because the lower hot ductility. There would be segregation-enriched at the cracks, which take the unfavorable influence for the heat treatment properties of steel.

5.5.2.2 Cause of rhomboidity

Four corners shell of the billet cooled uneven at the cooling beginning caused the rhomboidity. The diagonal two angles which cooled strongly would become acute angles; the counterparts would become obtuse angles.

5.5.2.3 The methods to reduce rhomboidity

(1) Use the two taper mould and the parabola type mould.
  (2) Narrow water seam and high mihaya of the mould.
  (3) Decrease the mould level fluctuation.
  (4) Keep the precision of subtended arc.
  (5) Ensure cooling even in secondary cooling zone.
  (6) Liquid steel composition controlling.

5.5.3 Deformation of round bloom

5.5.3.1 The reasons caused round bloom deformation

The round bloom could become to be elliptic or irregular polygon. The bigger cross-sectional diameter is, the more seriously the round bloom deformed. The reasons caused round bloom deformation are the following:

(1) The internal shape of mould has been deformed.

(2) The cooling is not uniform in secondary cooling zone.

(3) The arc alignment precision of continuous caster lower part is bad.

(4) The adjustment of clamping force to pulling-straightening roller is not suitable and without soft reduction.

5.5.3.2 The corresponding measures to reduce the round bloom deformation

(1) Change the mould deformed in time.

(2) Technique to guarantee the arc alignment precision of caster.

(3) Cooling uniform in secondary cooling zone.

(4) Decrease the casting speed properly.

5.6 Breakout

Breakout is the most detrimental event associated with the continuous caster. When a breakout happens, the solid strand shell tears such that a loss of liquid steel containment takes place and the liquid steel pours out beneath the mould. These breakouts are of a significant concern in the steelmaking industry, because they can:

(1) Cause severe downtime leading to reduced production.

(2) Create substantial costs due to repairing or replacing damaged equipment.

(3) And most importantly, because they pose a significant safety risk to plant operators. It was reported that a typical breakout for a conventional slab caster can result in costs approaching US$ 200,000. Therefore, it is necessary to reduce breakout occurrences for successfully achieving stable operation of the continuous casting process.

A breakout may occur either during the start-cast operation referred to as a start-cast breakout, or during the following run-time operation referred to as a run-cast breakout. Dofasco's plant experiences show that approximately 25% of all breakouts actually fall into the start-cast breakout category. Both start-cast and run-cast breakouts can be avoided to some extent by reducing the casting speed thereby providing more residence time in the mould for the steel to solidify. In order to avoid the occurrence of a breakout, it is crucial to detect improper solidification of the steel shell in advance with enough lead-time to appropriately slow down the caster.

5.6.1 Cause

(1) Casting temperature (superheat).

If the temperature of casting is too high, the steel shell would be thin. The friction between shell and copper plate, which would be higher because of the static pressure of mould powder and steel, caused the bonding.

(2) Performance of mould powder.

Bad lubrication performance of mould powder is the main reason caused the breakout. The parameter of mould powder includes melting temperature, melting rate, crystallization temperature, solidification temperature and viscosity, etc. The mould powder has good lubrication performance which make the process be fluent when the casting temperature was appropriate; if the casting temperature is too high or too low which made the lubrication bad, the friction between shell and copper plate would be higher, it may cause the bonding.

(3) Composition of steel.

Carbon is not only the basic element of steel but also the most effective element to the crystallization structure. The heat flow is least when the carbon content of steel is about 0.12%. The phase transition of shell here cause the volume shrinkage, which make an air gap between the shell and the mould, the heat flow is least and the shell thin and uneven. Because of the static pressure of steel, the cracks even the breakout may occurs when the out of the mould exit.

(4) Casting speed.

In a normal speed condition, the casting speed, the supply rate of mould powder and the cooling water of mould copper plate are stable. When the casting speed changed suddenly, the temperature fluctuation of mould copper and the discontinuous supply of mould powder would occurs result in the shell direct contact with the copper plate. It may lead the breakout.

5.6.2 The methods to prevent the breakout

(1) Select the mould powder with good performance.

    (2) Ensure the process operation is stable.

    (3) The precision of mould is good.

    (4) Good oscillation mode for continuous casting machine's mould.

    (5) Real-time monitoring prediction system.

5.6.3 Application of multivariate PCA for breakout prevention

A number of approaches have been proposed to predict breakouts in continuous casting processes, including:

(1) Pattern-matching methods(Yamamoto, Kiriu, Tsuneoka & Sudo, 1985; Nakamura, Kodaira, & Higuchi, 1996).

(2) MPCA(Multiway Principal Component Analysis) methods. The sticker detection method is one of the typical pattern-matching methods, which are comprised of comprehensive rules to characterize patterns in the mould temperatures(measured by thermocouples located around the mould) prior to the incidence of a breakout based on past casting operation experiences. MSPC methods have also been applied to breakout prediction in the continuous, run-time operation(Vaculik et

al. ,2001), where a PCA model is built using selected process measurements to model the normal operation of casting processes; certain statistics are then calculated by the model to detect abnormal operations including breakouts.

Both of these methods, unfortunately, cannot be employed to detect and prevent a breakout due to some limitations of the applied technology. Therefore, there exists a strong need to be able to real-time online monitor the cast and detect impending cast breakouts with sufficient lead-time such that they can be prevented by taking appropriate control actions. One possible control action is to change the ramping profile of the casting speed in order to slow down the casting process and provide more time for steel solidification in the mould.

5.7 Effect of Spray Cooling on the Quality of Continuous Casting Semis

The mould exit temperature and the cooling rate, in addition to the mould-cooler distance are the most important factors which affect positioning the solidification front and consequently the occurrence of some of the previously explained cast surface defects. Maintaining the mould exit temperature just above the solidification temperature of the melt is the most crucial factor. The cooling rate has to be increased properly to meet the demand for cooling the cast. Increasing the cooling rate is mainly done by increasing the cooling water flow rate and/or using multiple cooling units in order to increase the cast surface area at which heat is extracted. Using excessive cooling unit length in the case of the bath cooling device is equivalent to the effect of using multiple units of smaller lengths. For example, it was found that the bath cooling device length of 10cm exceeds the required length of approximately 1.5cm for cooling the cast down to room temperature during casting. Advantageously, the excessive cooler length of 8.5cm represents an automated and reserved potential for cooling when needed. This leads to a subjective conclusion that excessive cooler length, may be an effective factor in replacing the need for increasing the cooling water flow rate, or reducing the mould-cooler distance as an automated compensation for the demand of increasing the cooling rate that results from an increase in the casting speed.

One of the most severe defects in continuous casting products is concerned with the cracks provoked by improper design of the spray cooling system.

Below the mould, pressure from the liquid steel within the solidified shell can give rise to inter roll bulging, resulting in strains at the solidification front, which can cause cracks formation, and which is penetrated with solute enriched liquid(segregation). Prevention of this type of crack is achieved by a well set up spray system to produce a shell that can withstand the Ferro-static pressure. This shell behavior is related to surface temperatures below an upper limit as 1200℃ for billet casting.

Halfway cracks are related to billet surface reheating along the caster. The reheating is the result of a sudden decrease in the heat extraction rate from the surface as the billet moves from the mould to the sprays, from a spray zone to the next one or from the sprays to the radiation cooling zone. As a consequence, the surface tends to expand and tensile stresses at the solidification front develop provoking hot tearing. The amount of permitted reheating which avoid halfway cracks depends on a

5.7 Effect of Spray Cooling on the Quality of Continuous Casting Semis

number of factors. One of the most important factors is casting structure. Steel with a predominantly equiaxed structure can withstand the crack formation until reheat of 200℃, while columnar structure is more conducive to cracks with lower surface reheating.

Some researchers worked to develop a two dimensional heat transfer model based on the finite difference method in order to calculate the strand temperatures and the solid shell profile along the machine. An Artificial Intelligence heuristic search procedure interacts with the numerical model to determine the improved cooling conditions for the sprays zones of a real continuous caster for the production of quality casts.

6 New Approaches in Continuous Casting of Steel

There are significant new achievements in the continuous casting of steel since 70's, such as mould level control and instrumented mould with thermocouples, electromagnetic stirring(EMS), tundish metallurgy, numeric of temperature and flow patterns, electromagnetic brake in mould(EMBr), thin slab casting, liquid core reduction(LCR), dynamic soft reduction, high speed casting, mould flow control(MFC), dynamic spray control, near net shape casting, the hot charging, direct strip casting, and the newly 3D online control. These innovations increased the production ability, improved the quality, helped the resource efficiency. In this chapter, we will introduce some of them respectively.

6.1 Metallurgical Techniques of Tundish

At the casting machine, two steps are involved in transferring the liquid steel from the ladle to the moulds. Liquid steel is first fed continuously or semi-continuously from the ladle to the tundish which, in turn, distributes the liquid steel through nozzles in a continuous flow to the individual moulds. Metal flow through the ladle nozzle into the tundish is controlled by a stopper rod mechanism or by hydraulically or electrically controlled slide-gate systems. (The latter systems are rapidly receiving acceptance because of greater control capabilities and reliability.)

So, the tundish should receive molten steel from the ladle and apportion it to the individual strands with the least possible heat loss, while preventing the re-oxidation of the molten steel and the passage of inclusions.

To produce high-quality steel products with high productivity, effective technologies for cleanliness in molten steel have been desired in the steelmaking process. Inclusion separation and slag removal are important for molten steel cleanliness. Therefore, the following technologies have been proposed and introduced contamination in the tundish:

(1) Use of a large capacity tundish to obtain a residence time long enough for inclusions to float.

(2) Improvement of the dam, apt to the control molten steel flow.

(3) Promotion of inclusion separation by heating the molten steel in the tundish.

(4) Use of Ar gas blowing into the molten steel of the tundish.

(5) Development of a process in which molten steel can be fed from two ladles during ladle exchanges.

(6) Scaling of the tundish to prevent oxidation by the air.

(7) The centrifugal flow tundish in which the molten steel is horizontally rotated by electromag-

netic force has been developed.

In this part, we will mainly introduce the heating technology of tundish, tundish flow control, tundish flux and the centrifugal flow tundish.

6.1.1 Heating technology of tundish

While a tundish is initially being filled, the bath temperature fall blow that required for quality. Since this temperature drop also causes troublesome nozzle clogging, a tundish equipped with heating device is a preferable.

The prime advantage of the high temperature tundish is that it allows a reduction of the tapping temperature for billet casters of up to 50℃, with consequent saving in steelmaking furnace refractory lining costs and lowered risks of phosphorus reversion. The latter makes it of particular interest when converting high phosphorus hot metal.

There are two types of heating device: induction heating and plasma heating. Either device ensures control of the bath temperature within 5℃ throughout the casting period. An example is shown in Fig. 6-1 which indicates that the plasma heating ensures a uniform bath temperature in the tundish. The heating device enables us to adjust the bath composition by adding small amounts of alloying elements into the tundish.

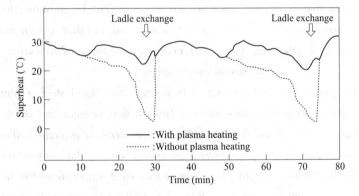

Fig. 6-1　Effect of plasma heating on the elimination of temperature drop during the ladle exchange

Drastic cost reduction is expected if the hot tundish is repaired and used repeatedly after the remaining slag and metal is dumped out. A drawback is, however, that the next heat is contaminated by the former slag remaining in the tundish; this was overcome by using some fluxes with high basicity and low viscosity. Fig. 6-2 illustrates how to reuse the hot tundish.

Moreover, mathematical modeling study has shown that the heat transfer and fluid flow characteristics of the melt in the tundish cannot be treated in isolation but have to be coupled with the heat transfer characteristics of the ladle for realistic simulation of the continuous casting process.

6.1.2 Flow control in tundish

The tundish in a continuous casting operation is an important link between the ladle, a batch vessel, and the casting mould with a continuous operation. It has traditionally served as a reservoir and

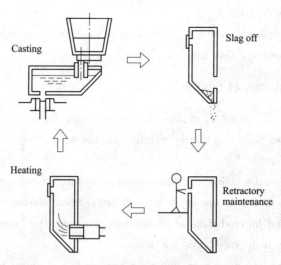

Fig. 6-2　Illustration of how to reuse the hot tundish

distributor of molten steel but now, its role has considerably expanded to deliver metal of desired cleanness and composition. So, in the continuous casting of steel, the task of the flow system is to transport molten steel at a desired flow rate from the ladle into the mould cavity and to deliver steel to the meniscus area that is neither too cold nor too turbulent. In addition, the flow conditions should minimize exposure to air, avoid the entrainment of slag or other foreign material, aid in the removal of inclusions into the slag layer, and encourage uniform solidification. Achieving these somewhat contradictory tasks needs careful optimization.

The tundish flow pattern should be designed to increase the liquid steel residence time, prevent "short circuiting" and promote inclusions removal. Tundish flow is controlled by its geometry, level, inlet(shroud)design and flow control devices such as impact pads, weirs, dams, baffles, and filters.

The tundish impact pad is an inexpensive flow control device that suppresses turbulence and prevents erosion of the tundish bottom where the molten steel stream from the ladle impinges the tundish. The incoming stream momentum is diffused and allows the naturally buoyancy of the warm incoming steel to avoid short circuiting, particularly at startup. Together with weir and dam, the TURBOSTOP pour pad improved steel cleanliness, especially during ladle exchanges. At Lukens Steel, T. O decreases from 26ppm(with a domed pad)to 22ppm(with a hubcap pad). At POSCO, steel cleanliness was improved by putting 77 holes in their dam, making it act as a partial filter. At Dofasco's #2 Melt Shop, using baffles improved product quality, especially at ladle exchanges, thereby making the heat more consistent. Baffles combined with an initial tundish cover lowered the average T. O. in tundish during steady state casting from 39±8ppm to 24±5ppm.

6.1.3　Tundish flux

The tundish flux must provide several functions. Firstly, it must insulate the molten steel both thermally(to prevent excessive heat loss)and chemically(to prevent air entrainment and reoxidation). For example, at Imexsa Steel(Mexico), by changing tundish flux(with lower SiO_2 content), nitrogen pickup from ladle to mould decreased from 16ppm to 5ppm.

Secondly, in ideal circumstances, the flux should also absorb inclusions to provide additional steel refining. A common tundish flux is burnt rice hulls, which is inexpensive, a good insulator, and provides good coverage without crusting. However, rice hulls are high in silica and (SiO_2 80%), which can be reduced to form a source of inclusion. They also are very dusty and with their high carbon content, (C 10%), may contaminate ultra low carbon steel.

Basic fluxes (CaO-Al_2O_3-SiO_2 based) are theoretically better than rice hulls at refining steels, and have been correlated with lower oxygen in the tundish. For example, the T.O decreased from 25-50ppm to 19-35ppm with flux basicity increasing from 0.83 to 11, measured in practical cases. At some melt shop, using basic tundish flux (CaO 40%, Al_2O_3 24%, MgO 18%, SiO_2 5%, Fe_2O_3 0.5%, C 8%), together with baffles, significantly lowered in total oxygen fluctuation, as compared to the initial flux (CaO 3%, Al_2O_3 10%-15%, MgO 3%, SiO_2 65%-75%, Fe_2O_3 2%-3%). The T.O decreased from 41ppm to 21ppm during ladle transitions and decreased from 39ppm to 19ppm during steady state casting.

However, other results, such as shown in Fig. 6-3 found no improvement in T.O between rice hulls and higher basicity flux (SiO_2 25.0%, Al_2O_3 10.0%, CaO 59.5%, MgO 3.5%). This might be because the basic flux still contained too much silica. More likely, the basic flux was ineffective because it easily forms a crust at the surface, owing to its faster melting rate and high crystallization temperature. Also, basic fluxes generally have lower viscosity, so are more easily entrained. To avoid these problems, it is suggested a two-layer flux, with a low-melting point basic flux on the bottom to absorb the inclusion, and a top layer of rice hulls to provide insulation, which lowered T.O from 22.4ppm to 16.4ppm.

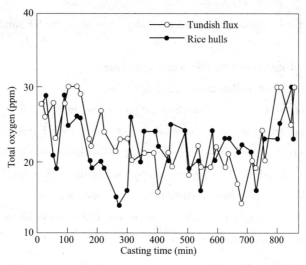

Fig. 6-3 Effect of tundish flux on the T.O in tundish

6.1.4 Centrifugal flow tundish

A new process was devised for promoting inclusion separation in a tundish. The process utilizes electromagnetic force to rotate molten steel in a cylindrical tundish. The centrifugal force caused by the rotational flow promotes separation of inclusions from the molten steel. This tundish is referred to as the centrifugal flow tundish (CF tundish).

The general concept of the CF tundish is shown in Fig. 6-4. A moving electromagnetic field is imposed from outside a cylindrical tundish. Molten steel is horizontally rotated and centripetal force acts on the inclusions due to their lower density relative to that of the molten steel. Inclusions are separated from the molten steel, and clean steel out from the bottom corner of the tundish into a mould.

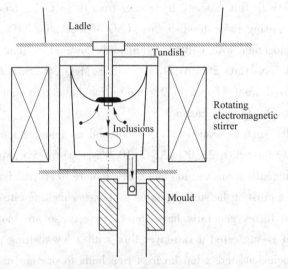

Fig. 6-4   Concept of CF tundish

Three mechanisms for inclusion separation and slag removal in the CF tundish were anticipated, as follows:

(1) Concentration of inclusions by the centripetal force.

(2) Promotion of inclusion collision and agglomeration.

(3) Improvement of the residence time distribution by the rotational flow.

Industrial plant tests carried out at Chiba Works showed that the CF tundish has excellent de-oxidation performance estimated at 0.17-0.25min^{-1} as a de-oxidation rate constantly. The centripetal force and large amount of turbulent energy caused by the rotational flow accelerates inclusion separation. Furthermore, the residence time distribution of the molten steel in the tundish is improved, and slag removal during the ladle exchange is promoted. The process is successfully used in the commercial production of high-quality stainless steel slabs.

6.2   The Mould-level Control

Mould-level control is show in Fig. 6-5.

The most vital part of the control of a continuous casting machine is to ensure that the withdrawal of the cast and the partially-cooled billet is such as to keep the liquid level in the

6.2 The Mould-level Control

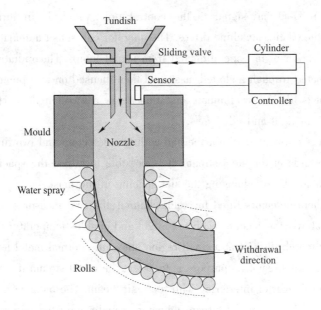

Fig. 6-5 Mould-level control in continuous casting process

mould constant(within a few centimeters). This is done in two ways:

(1) The tundish is weighed and the rate of feed to the tundish from the ladle varied automatically to keep the total tundish weight constant. In this way the rate of feed from the tundish is constant.

(2) The rate of withdrawal of the partially cooled billet is controlled so as to keep the level of liquid steel in the mould roughly constant.

In the early days of continuous casting the level of the top of the liquid steel in the caster was maintained constant by an operator viewing it and adjusting the tundish stopper accordingly. It is now normal to have a means of finding the level using a measuring instrument and automatically adjusting the level. The Table 6-1 below lists several ways in which the level is detected. Two of them, the gamma-ray(radioactive) and the infrared methods will be described in detail.

Table 6-1 **Ways of the level**

Type	Maker	Comments
Gamma-ray	Distingtan Engineering(UK)	Widely used, reliable
Eddy-current	NKK(Japan)	
Thermistor	United States Steel	Only on Uss machine
Infrared	Sert. Danielli	Widely used on the continent of Europe
Electro-magnetic coil	Concast	

The operation is self-evident from this diagram. The infrared device was developed in order to avoid the use of powerful radioactive isotopes. The detector views the junction of the metal level with the back wall of the mould. As the metal level rises within the field of view more radiation is received by the single photocell and an increased output is obtained. Special provisions are made to compensate for interruption of the view of the metal. The photocell unit receives the infrared radiation and provides an electrical signal to the control unit, which is in turn connected to the operator's unit and the casting-machine drives. The operator can select automatic or manual control and he receives indication of the operating rod from signal lamps. The radiation emitted from the liquid steel is collimated through a slotted mask and then focused on to a photodetector by a cylindrical lens. The light is filtered to eliminate radiation below a wavelength of 1mm, so reducing interference from ambient light and oil flames.

The entire system is duplicated within the had with two detectors and two fit beams normally arranged to view either side of the steel stream. It is possible to adjust the spacing between the two areas seen by the photocells by changing the slot spacing in the mask.

There are three photo detectors fitted for each channel: the first measures the metal level using the beam described above; the second receives to light and enables temperature drift compensation; and the third looks through the slot at a small region above the normal metal level and between the main beam and the metal stream. Its purpose is to detect the metal stream if it wanders from a central position and is in danger of interfering with the main beam. The balance between the two main beams and the threshold level of the stream detectors can be adjusted with small potentiometers mounted in the back of the unit.

The level signal detected by each channel is fed, after temperature compensation, to a simple circuit which selects the largest signal. Thus, the unit always controls on the higher of the two-level signals. If the stream-sensing photocell sees that the teeming stream is moving towards the detection beam it blocks the signal and the unit switches to control on the other channel. There is an additional feature that if both channels are blocked together, for example by a fan-shaped metal stream, the unit switches to a memory, equivalent to the fast detected metal level, and prevents a sudden loss of control. As the memory decays the metal level gradually drops allowing the operator ample time to intervene.

The unit gives a smooth transition from manual to automatic control by preventing automatic operation if there would be a large jump in withdrawal speed at changeover. It does not provide bumpless transfer when changing from automatic to manual. There is also protection against changing to automatic when there is a cable fault.

The control system receives the chose level signal and, following proportional and integral action. Outputs a voltage signal directly to the withdrawal drive unit. The drive creates a withdrawal speed proportional to this voltage signal.

6.3 Hot Charging and Direct Rolling of Continuous Casting Slab

Although the practice of hot charging a semi-finished shape into the reheating furnace of the finishing mills is not necessarily a productivity improvement attributable to continuous casting, it is, nevertheless, receiving wide attention because of the potential fuel saving.

In the early development of continuous casting the product was cooled to ambient temperature, inspected for defects and, if necessary, conditioned to remove the surface defects (a practice that is comparable to that used for many ingot-rolled, semi-finished products). The product was then reheated and further processed in the finishing mills which is energy and labor intensive method. By charging hot continuously cast product into the finishing mills, the sensible heat of the product is utilized with significant energy savings. This practice may avoid reheating altogether or require some intermediate reheating. However, it demands close coordination between the caster and finishing area. It also demands excellent surface quality because on-line hot inspection and conditioning of the cast material is not yet fully developed. The main technological process is shown in Fig. 6-6.

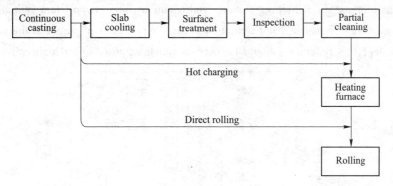

Fig. 6-6 Comparison of hot charging, direct rolling and conventional technological process

6.4 Soft Reduction at the Final Stage of Solidification

It is well known, that centerline shrinkage and segregation of continuously cast strand are the main defects. The centerline segregation in CC steels is known to be formed by inter-dendritic liquid flow at the final stage of solidification of CC strand. Major factors which result in inter-dendritic liquid flow are the solidification shrinkage and the bulging of solidified shell. Along with the increasing demand for high quality steels, elimination of segregates even as small as several millimeters in radius has become one of the major subjects for CC engineers.

The soft reduction applied at the final stage of solidification has obvious effect on internal quality. The technology of soft reduction at near solidification end originated between the late 1970s and early 1980s. Firstly, the soft reduction technique was put in use and successfully at Nippon Steel in Japan in 1976. In 1986 the soft reduction unit of improved design was installed in No. 4 slab caster in Fukuyama Works. At present, it is applied worldwide on slab and bloom casters, and regarded as the best choice for further improving the macro-segregation of a bloom. As an important measure to improve the product quality and to develop high-profit products, the soft reduction technique has become the constitutional part of a modern continuous caster. Many quantitative models based on thermal-stress model and shrinkage criterion function are proposed, which are able to estimate the effects and position of soft reduction more appropriately, implement the technique more scientifically.

6.4.1 Effects of soft reduction

(1) Improvement in macrostructure.

The soft reduction can obviously improve the macrostructure due to its compensation for solidification shrinkages during casting. The frequency of central segregation, central porosity, and looseness rating over 1 class decreased obviously, while the frequency of 1 class and lower increased greatly.

The more the reduction amount, the better the results are (see Fig. 6-7). As for high carbon rail steel, because of the larger solidification shrinkage ratio, the macrostructure was consistently improved with an increase in reduction amount from 3mm to 7mm; even 7mm reduction, no inner cracking was identified. Because 7mm reduction is the limit of the equipment at the normal casting speed for rail steel, this reduction should be used as much as possible for high carbon steel.

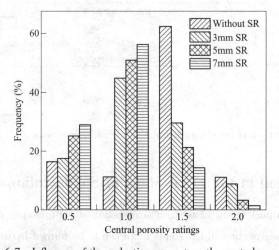

Fig. 6-7 Influence of the reduction amount on the central porosity

The effect of the location and length of reduction region on the central segregation is far larger than that of the reduction amount, so the casting parameters affecting the location of solidification end can significantly influence the metallurgical results of soft reduction. The variation of casting parameters changes the length of reduction region and maximum reduction amount; for example, at a casting speed of 0.75m/min, 5 frames can carry out soft reduction, and the maximum reduction amount can reach 7mm, but for a casting speed of 0.7m/m, only 4 frames can carry out reduction, and the maximum reduction amount decrease to 5mm, and the metallurgical results of soft reduction is reduced.

(2) Improvement in composition uniformity.

Soft reduction has influences on the transversal and longitudinal carbon distributions of blooms. Fig. 6-8 illustrates the influence of soft reduction on longitudinal central carbon segregation. With soft reduction, the fluctuation of the central carbon segregation index C/C_0 (C and C_0 are the composition of solid and liquid steel, respectively) is decreased greatly compared with that of blooms without reduction.

As for U71Mn bloom, when the casting conditions are superheat = 31℃, v_c = 0.70m/min and 5mm soft reduction, the carbon segregation indexes are 0.93-1.08 and 0.96-1.03 for the blooms with and without soft reduction, respectively. The serious carbon segregation may induce the forma-

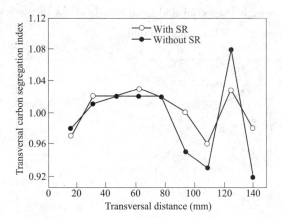

Fig. 6-8 Influence of dynamic SR on the transversal carbon segregation index

tion of the brittle phase such as cementite and martensite in high carbon steel. Moreover, by examining the chemical composition distribution of rails, Guijun Li et al found that the range of composition deviation between two random points is as follow (wt%): C, 0.02; Mn, 0.05; Si, 0.02; P, 0.005; S, 0.005; and the requirements of high-speed rails for 350km/h can be met.

(3) Mechanical properties.

Until April 30, 2004, the rail and beam plant of Pangang had produced 354,500t of heavy rails from continuously cast blooms. According to the statistics, the mechanical properties of heavy rails are excellent. Due to the precise composition control and the less segregation obtained by soft reduction, the tensile strength and toughness were within a narrow region, for example, for PD3, the tensile strength was 1030-1060MPa (the average value was 1045MPa), the elongation was 10.5%-12.0% (average 11.4%). In the standard falling weight test, no fracture was obtained, and the deflection of the test sample was within 37-44mm (average 39mm); all the properties met the requirements of high-speed rails for 350km/h.

6.4.2 Problem of attention in soft reduction

The above analysis concludes that soft reduction is an effective method to eliminate the center segregation and center porosity of billets. However, soft reduction also has unfavorable influence on the internal quality of billets, i.e. internal cracks in billets, if soft reduction is conducted at an unsuitable time, namely, with unsuitable solid fraction. These cracks are located in the columnar crystal zone over the longitudinal section of billets, perpendicular to casting direction and filled with segregated residual molten steel. The effect of the waiting time and solid fraction on the internal cracks is shown in Fig. 6-9. As shown, the internal crack index decreases from 50 to 0mm as the waiting time rises from 180s to 270s and as the solid fraction in the core of billets increases from 0.65-0.78 to 0.96-1.0. This is reasonable because the solidified shell becomes thicker as the waiting time increases. Furthermore, at waiting times longer than 270s, the solid fraction becomes larger than 0.96. At such a large solid fraction, the cooling rate at the crater tip is higher than that on the surface of the billets and the volume contraction in the core area is larger. As a result, the macro center cracks will form through tangential and axial tensile stress in the core area if the shrinkage in solidification in the core area is not compensated.

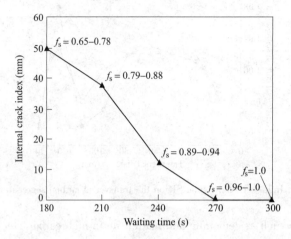

Fig. 6-9 Relation between the waiting time/solid fraction and inter crack index

By conducting tests on soft reduction employing single and multiple pairs of one-piece rolls, the influence of the amount of reduction and reduction timing on center segregation were studied. They found that center segregation was improved with increasing amount of reduction in the case when soft reduction was applied only in the final stage of solidification where the fraction of solid at the slab center was greater than 0.25. At the same time, variations in the casting direction and the distribution of center segregation in the transverse direction of the slab were also improved. The deteriorating tendency of center segregation due to soft reduction conducted in the wide stage of solidification stage was considered to be attributable to the new flow of molten steel caused by soft reduction. For the improvement of center segregation, it is considered necessary to optimize the reduction timing and to suppress uneven reduction due to mechanical factors such as roll bending.

From the above analysis, it is clear that by applying soft reduction, the internal cracks, macro cracks, and the center segregation must be considered at the same time. When the solid fraction is very small, the internal cracks will be easily induced by soft reduction. In contrary, if the solid fraction is very large, the center segregation cannot be effectively eliminated and macro cracks will appear in the billets.

6.5　Electromagnetic Stirring Techniques

Though continuous casting can provides higher yield, uniform product of a good quality, and higher overall productivity, there still remain some considerable quality problems, which mainly appear in three zones namely surface, subsurface and centerline. A successful solution has been found, that is the application of electromagnetic stirring.

6.5.1　Historical development of electromagnetic stirring

Although electromagnetic stirring technique found its application in continuous casting process only

6.5 Electromagnetic Stirring Techniques

in recent years, the idea of stirring a metal during its solidification in an attempt to more accurately control the process was being studied before the advent of commercial continuous casting.

The first electromagnetic stirring device specifically adapted to continuous casting was designed in 1952. The device was located beneath the mould. The first trial on in-mould stirring was conducted in Kapfenberg at Boehler. Line frequency (50Hz) was used in this case to guarantee penetration of the 50Hz rotating magnetic field into the mould, low conductivity mould materials (aluminum bronze or molybdenum) had to be used. The experiment showed improvement on the surface quality of the cast product. By the late 1960's, work on below-mould stirring had been conducted on billet and bloom casters in France, UK, USA., and USSR. Both rotating stirrer and liner stirrer had been investigated by this time. All these early investigations confirmed substantial improvements on the metallurgical quality of the cast products. Also from this work, the appearance of "white band" was found. Suggestions to eliminate "white band" were given such as using a very mild stirring over a large length of the strand.

The first successful industrial application of electromagnetic stirring in steelmaking goes back to 1973 by IRSID/CEM on a 240mm 2 bloom caster at SAFE in France. From then on, the application of electromagnetic stirring on continuous casting machines rapidly spread worldwide.

The Japanese steel industry began to develop electromagnetic stirring only at this point of time but soon came to a leading position in this field. By now they have adopt electromagnetic stirring on a very large scale on continuous casters.

Various electromagnetic stirring devices were first designed for below-mould stirring. These include the rotary stirrer, the cylindrical linear stirrer, the helicoidal stirrer and the flat linear stirrer. Improvements on centerline segregation, porosity and V-segregation were reported.

Development of in-mould stirring was first used by IRSID and later by BSC. Low frequency was used in these cases so that the traveling magnetic field could penetrate through the conventional copper mould. Improvements on surface and subsurface qualities were reported.

Later on rotary in-mould stirring was used and the report claimed that, in addition to the benefits obtained in surface and subsurface qualities, the centerline quality was also improved.

There were also some unusual types of in-mould stirrers reported. These include so called "electromagnetic brake" for slab caster, the main frequency mould stirrer, and the permanent magnet stirrer which was driven by a hydraulic motor.

By the mid-70's, after wide industrial use, the limitations of a single electromagnetic stirring stirrer showed. Satisfactory quality demanded combination stirring, i.e. the use of several stirrers at different locations of the same strand, such as in-mould, at the secondary cooling zone and at the end of the solidification zone.

The development of electromagnetic stirring for slab casters lagged a few years behind the billet and bloom casters due to the complexity of the problem. Work here also started from below-mould stirring. Both inductive stirrer and conductive stirrer were used.

In-mould stirring of slab casters was initiated by IRSID in 1978, where linear vertical stirrers were used. Later on horizontal linear stirrers were developed.

It is only a few years since the first electromagnetic stirring device was installed on the production strand of continuous casting. But today, hundreds, if not more, of electromagnetic stirring machines have been implemented worldwide.

6.5.2 Effects of electromagnetic stirring

Electromagnetic fields can be used to create strong, circulating flow patterns in metallic liquids. Theoretical methods of estimating flows in such systems have been developed the potential benefits to be derived from electromagnetic stirring of liquid steel during solidification are receiving wide attention. Improvements reported include:

(1) Internal quality(reduced segregation, cracking and porosity) through a preferred solidification structure.

(2) Sub-surface and internal cleanliness through a modified metal flow pattern.

(3) Reduced criticality of casting parameters(temperature and casting speed).

(4) Increased productivity through increased casting speeds.

Electromagnetic stirring enables the operator either to improve the quality of cast billets while keeping the casting speed constant or to increase the casting speed without impairing the quality of the product.

Electromagnetic stirring to improve the as-cast structure of semis by the method developed by IRSID is used at a French steel plant where a four-strand bloom caster is equipped with stirrers beneath the mould. Fewer defects on finished products and improved fatigue properties of automotive forgings have resulted.

Tests in the USA have shown similar grains, but the improvement was not enough to offset the operating difficulties of fitting stirrers close to the mould.

For high carbon wire of small diameter, Japanese studies have shown that the stirred steel gives a much lower percentage of breakages during drawing than the non-stirred steel.

In England upwards movement of liquid metal along the solidifying surface of the strand, induced by electromagnetic stirring, has been tested in a search for improved surface and lower segregation, using a stirrer in the mould.

Today, there is practically no quality billet/bloom caster in the world that has not adopted this process. The same may be said of stainless steel slab casters. Since the 1980s, they have recognized the benefits of electromagnetic stirring for equiaxed zone or grain size control. For slab casting, the application of electromagnetic stirring is more recent. The aim is the same, that is to say to break the dendrites along the solidification line and to decrease the width of the columnar zone. But due to the width of the slab, this effect is more difficult to achieve than it is for blooms or billets.

For low carbon steels cast on conventional thick slab casters, however, electromagnetic stirring applications remain limited. DC brakes were introduced by KSC and ABB in 1982. They were tested and implemented on several curved-mould machines with some success but really did not break through has initially projected(Vertical/bending remains the best way of offsetting gravity).

Rotary stirring, induced by two long linear stirrers placed at the meniscus, was introduced by Nippon Steel at about the same time to address shell formation and subsurface slab quality. To improve the centerline segregation, intensive studies on the electromagnetic stirring (EMS) have been carried out. However, EMS cannot be applied to the slab caster successfully because the packing of equiaxed crystals without segregation is found to be very sensitive to the flow pattern induced by EMS at near the crater end. The EMS application is also limited to only medium carbon steel because of the difficulty of crystal multiplication in other steels.

Plates made from stirred slab have less central segregation and thus present improved welding properties.

6.5.3 Types of electromagnetic stirrers

Electromagnetic stirrers were initially installed on billets casters to reduce centerline segregation. This was achieved by a change in solidification structure: the area of the central equiaxed crystal zone was increased with a corresponding decrease in the area of the outer columnar crystal zone. Subsequently, the other improvements listed were recognized. In a recent survey, it was reported that over 100 stirrers are in operation, of which over 60 are on billet and bloom machines with approximately 40 on slab casters. There are two basic types of stirrers, (rotary and linear) which can be installed either in or below the mould.

(1) Rotary stirrer.

As shown in Fig. 6-10, in a rotary system installed in the mould of a billet caster, a rotating magnetic field produced by the coils imparts a circular motion to the liquid steel. The centrifugal force developed results in a sound skin, with the lighter phases (i.e., inclusions) moving towards the center. The central equiaxed zone is enlarged because the rotational flow promotes the fracturing of the tips of the columnar dendrites which serve as nuclei for equiaxed crystal formation in the central zone.

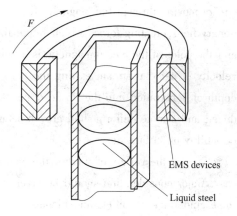

Fig. 6-10 Rotary stirrer

In a continuously cast billet, a rotating field can initiate rotation of the metal in the liquid pool and in a continuously cast slab, liner fields are used to move the liquid metal along horizontal or vertical axes.

In billet casting, the detailed technology can differ but, generally, the electromagnetic stirrer is a water-cooled ring with induction coils surrounding the strand which are positioned at a level depending upon the billet size and the casting speed. In the case of the stirrer developed by the institute de Recherches de la Siderurgie Francaise (IRSID) and Compagnie Electro-Mecanique (CEM) in France, the diameter of the stirrer is about 0.5m, the distance between the coil and the bottom of the mould varies from 2 to 4m, according to the size of the billet, the frequency 50Hz and the power usually between 1 and 5kVA per ton per hour.

(2) Linear stirrers.

With the linear system, electromagnetic coils are installed along the side of a strand (below the mould) which produce a vertical circulation pattern. The increase in the central equiaxed crystal zone is obtained by a similar mechanism as that obtained by the rotary stirrers. Inclusions, which are normally concentrated in a head close to the upper surface in curved mould machines, are more uniformly distributed.

There are at least two kinds of technologies for the electromagnetic stirrer for the slab caster but their details are presently confidential. In Japan, and more particularly at Nippon Steel Corporation, the stirrer is made of two linear motors located in a water-cooled case placed close to the strand, with small auxiliary rolls pressing on to the strand. In the case 500kVA units provides a periodic pulse to agitate and free the growing dendrites, causing them to sink.

At Dillinger Huttenwerke AG tests have been carried out with stirrers developed by IRSID made of four linear motors placed inside two pairs of rolls of non-magnetic steel of exactly the same external size as the two pairs of conventional rolls they replace.

In another application of electromagnetic principles on a slab caster, a magnetic field is used as a break to modify flow patterns within the mould in certain areas and create flow patterns in others, with a subsequent improvement both in internal cleanliness and surface quality. This effect is achieved through the interaction of a moving steel stream in a stationary magnetic field. The metal stream moving through the magnetic field produces induced currents which, together with the stationary field, creating forces which brake the steel streams. In addition, steel between the streams and the poles of the electromagnetic is accelerated which provides a strong stirring action. Thus, the velocity of a metal stream exiting the ports of a refractory tube shroud is reduced as well as the depth of penetration into the liquid crater. Under these conditions, inclusion concentrations are reduced and a more uniform shell growth occurs around the periphery of the mould which lessens the possibility of surface defects.

So, among the variety of stirrers, the choice of a proper stirrer for a specific casting process is not easy. Major consideration should be given to metallurgical objectives. Based on this, Birat prepared some guidelines for billets and blooms casting, which are shown in Table 6-2.

Table 6-2 Guidelines for choice of optimum EMS for billet & blooms

Stirrer		Metallurgical objective		
Position	Type	Surface quality	Subsurface quality	Centerline quality
In-mould	Rotary	Yes	Yes	Yes
	Linear	?	Yes	?
Below-mould	Rotary	No	No	Yes
	Linear	No	No	Yes

Although various stirrers have been proposed and are in use, economical and efficient electromagnetic stirring devices are yet to be developed. There is a lot of work to be done in the electrical engineering side.

6.5.4 Metallurgical aspects of electromagnetic stirring

There have been some considerable quality problems in the continuous casting process. Three zones of cast strand have particular defects which H. S. Marr summarized in Table 6-3.

Table 6-3 Common defects in continuous casting

Region	Problem
Surface	Pinholes Slag patches Laps/double skin Reciprocation marks
Subsurface	Blowholes Inclusions (structure) (segregation)
Centre	Porosity Segregation Structure (inclusions)

Bearing in mind the problems shown in Table 6-2, the following attempts to explain how electromagnetic stirring has been applied as a remedy.

Pinholes are exposed small blowholes arising from dissolved gases emanating from solution as during surface solidification. Blowholes are larger bubbles trapped below the surface. Electromagnetic stirring, by creating a flow over the solidification front, effectively prevents the attachment of bubbles there.

Slag patches arise not only when refractory erosion or slag enters the mould, but also when the de-oxidation products float out or atmospheric oxidation of the steel occurs. Electromagnetic stirring can promote a flow pattern that guides the non-metallic inclusions towards sites of elimination. Particularly in the case of rotary mould stirring, centripetal forces draw the non-metallic inclusions away from the solidification front. Non-metallic inclusions can be entrapped by the growing dendrites. The risk is reduced by the strong convection velocities of liquid steel created by electromagnetic stirring. This so-called "wash effect" explains the clean subsurface obtained by all types of in-mould stirring.

Surface and subsurface qualities can only be improved by in-mould stirring, because these regions are already solidified when the strand emerges from the mould.

Severe problems occur on the centre axis or plane where the solidification fronts meet. One of the major problems has been the intensified degree of macro-segregation in the form of centerline segregation and associated V-segregation. Macro-segregation is the result of solute rejection at the solid-liquid interface followed by the mass transfer of this solute-enriched liquid along the interdendritic channel in the solid and liquid mushy zone. There are two different modes of solidification, i.e., columnar and equiaxed. The mushy zone is produced during the columnar mode of solidification. Therefore, any condition which effects an early transition from columnar to equiaxed growth will decrease the extent of macro-segregation and improve the quality of the continuously cast steel. Almost every report in this area has claimed an increase in the size of the equiaxed zone by using any types of below-mould stirrer and in-mould rotary stirrers. Before the application of electromagnetic stirring, the segregation problem was controlled by using a low thermal gradient, a low superheat and a low casting speed. This approach is not sufficient for higher quality steels and also severely limits process flexibility. The application of electromagnetic stirring relaxes these operational constraints and hence increases productivity.

Central porosity is another centerline defect, which is affected by changes in cross-sectional dimensions and in the amount of solidification shrinkage, the flow created by electromagnetic stirring will eliminate the formation of bridges, smooth out the solidification front, promote effective feeding from the liquid core higher up the strand to accommodate the shrinkage and hence suppress the formation of porosities.

Mechanical properties of stirred and unstirred materials have been compared. Encouraging improvements were claimed from tensile, fatigue and toughness teats, etc.

A relatively new but rapidly advancing steel casting technology is horizontal continuous casting. There are two major problems in spite of its many merits. Namely the gravitational effects and the inability of the process in produce wide slabs. Both these problems can be attacked successfully with the use of electromagnetic stirring.

The major drawback associated with the electromagnetic stirring process is the appearance of "white band". This is a result of inverse segregation and is so called because it appears as much on sulfur print or etched samples. The problem is more evident when a single powerful stirrer is used. Suggestions to alleviate the problem include limiting the stirring power, combination stirring and three-dimensional stirring.

As having been explained in this section, the application of electromagnetic stirring during the continuous casting of steel provides solutions to all the problems shown in Table 6-3. On the basis of the results reported in the literature, one should be confident of the substantial improvements in the qualities and properties of the cast products obtained with electromagnetic stirring. This technique has helped turn continuous casting into a process capable of manufacturing very high-quality steel and will play an increasing important role in this industry.

7 Special Continuous Casting Processes

7.1 Horizontal Casting

Conventional continuous casting has progressed from the totally vertical caster to the low head/ multi point straightening design with the major advantage in reducing overall caster height and ferrostatic pressure. The ferrostatic pressure for the vertical casters impose a severe duty on the support rollers and segments. Much development time has been concentrated on achieving virtually zero ferrostatic pressure by casting entirely horizontally. However, this involves going from a vertical feed into the mould to a device which allows a horizontal feed into the mould. This requires a horizontal tundish/mould joint and special conditions to reduce mould friction since the mould is rigidly fixed to the tundish by this feeding joint.

The mould/tundish link is made by a piece of refractory material which is called the break ring. The arrangement of tundish, nozzle, break ring and mould varies slightly depending on the machine builder. The break ring is made from a special refractory such as boron nitride or silicon nitride (Si_3N_4) which must be resistant to thermal shock, erosion and not be wetted by steel. Moreover, this part has to be machined to very accurate dimensions. Fig. 7-1 shows a cross-section diagram of a horizontal caster with the tundish and mould rigidly fixed. The horizontal casting process has the following advantages:

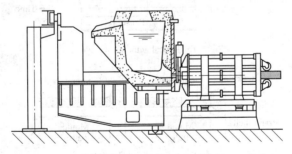

Fig. 7-1 Horizontal caster with stationary mould and movable tundish in casting position

(1) A very low head machine which can be installed in normal, existing buildings.

(2) The machine design gives full protection against atmospheric contamination particularly for small sections providing the capability of casting aluminium killed steels in small sections.

(3) The strand undergoes no deformation, which suits special grades such as tool steels and high alloy steels.

This is by no means a new process, the first experimental plants for steel casting date back to 1966-1967(Davy-Loewy in Great Britain, General Motors in the USA) but it has raised a lot of revised interest in recent years and much research and development work has been carried out to try to bring the process towards industrial application. There are approximately 30 plants which have been built since 1975; most of them remain pilot plants operated by the various machine builders but a few machines have reached the industrial production level, for example, NKK Fukuyama (1978), Boschgotthardshiitte (1980), Armco (1984), and British Steel (1988). Horizontal machines can be separated into two types according to their extraction mechanism:

(1) The mould tundish assembly is stationary and an intermittent extraction pattern is used. Fig. 7-2 shows (a) the tundish mould arrangement, (b) the typical withdrawal cycle and (c) the formation of the strand shell. This is the most commonly applied design and Technica Guss and Nippon Kokan/Davy Loewy have probably supplied most machines to date.

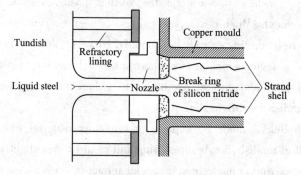

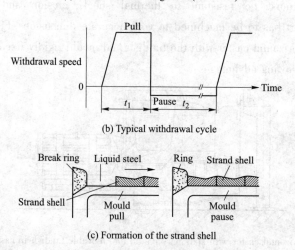

Fig. 7-2 Horizontal casting showing

(2) The mould tundish assembly is oscillating and a continuous extraction is used. This technique has been adopted in the USSR and by Krupp.

Two original developments should also be quoted: the Watts process which seems to have now been abandoned and a Russian development (VN Ⅱ Metmach) where one mould feeds two horizontally opposed strands simultaneously.

The sizes cast using horizontal casting are:

(1) Wires 3-12mm dia. on a 12-strand machine could produce 25,000 tpa.

(2) Rounds up to 330mm diameter.

(3) Billets and blooms 50mm square to 250mm square, 130mm×170mm.

Limitations with the break ring technique have so far not permitted slab casting nor big bloom casting. Casting speeds are similar to those achieved on a conventional machine but speeds are likely to increase with intensive development. The mould is not lubricated during casting and is made of copper alloy with high erosion and thermal distortion resistance. There is often a graphite section at the exit of the mould.

The main problems met with these processes are:

(1) Rounds up to 330mm diameter.

(2) The life and cost of the break ring which limits the casting time to a few hours, generally.

(3) The surface quality which must be free of transverse cracks associated with solidification marks, (also termed cold shuts or witness marks) which may require surface conditioning.

(4) It is difficult to supply lubricant into the mould to reduce high friction.

(5) It is not applicable to large section sizes with steel even though similar processes for nonferrous casting are fairly well developed.

The formation of the solidification marks results from the strand formation process at the break ring in the mould as shown in Fig. 7-2(c). Technological developments such as mould EMS, break ring shape, runner brick shape, oscillation or pulse frequency and an improved knowledge of the process (superheat, mould taper related to steel grade) has improved the surface quality. For optimum surface quality it is necessary to adjust the oscillation cycle for different steel grades.

The internal quality, in terms of central segregation and porosity, is similar to conventionally cast products but on large section sizes the structure presents some asymmetry. Electromagnetic stirring in the secondary cooling zone is another development to improve the internal quality which also eliminates the asymmetrical structure.

Initial application of the process has mainly been for special steels where the yield advantage compared to the ingot route is substantial but the tonnage requirements and capital costs moderate. For wider application break ring developments are crucial and the cost and service life of the break ring is probably the main limitation. This limits machine productivity and dominates the economics of the process. Whilst high quality refractory materials are developing rapidly considerable progress is still needed in refractory technology before the process becomes more widely accepted.

7.2　Beam Blank Casting

The first beam blank continuous casting machine went into operation nearly 30 years ago. At that time the continuous casting process itself was in its infancy and many so called experts questioned if the process could, first of all, be cost effective and secondly, if the process could produce the required quality.

The continuous casting process has changed dramatically over the past 30 years and so has the philosophy in producing the steel products. Based on the initial beam blank casting trials by the

British Iron and Steel Research Association, the Algoma Steel Corporation, with support by (SMS) Concast USA, decided to install an operational beam blank caster. The initial design of the twin strand beam blank continuous caster was based on five different beam blanks covering the required range of finished beams. The success of the project set the example for the production of steel products using a near-net-shape semi-finished section.

The success of producing wide flange beams by continuous casting so called "dog-bones" was quickly recognized by the Japanese steel industry in the early 1970's The table below (Table 7-1) generalizes the development of beam blank machines in Japan.

Table 7-1 Beam blank machines in Japan

Company	Date of start-up	Sizes (mm)
Kawasaki	1973	460×400×120
		560×287×120
Tokyo Steel	1979	445×280×110
Yamato-Kogyo	1980	460×370×140
NKK	1981	480×400×120

Kawasaki Steel ordered from (SMS) Concast the first large production machine. Designed in the United States and built in Japan, this caster advanced the full scale industrial application of the near-net-shape process.

The other Japanese companies producing beams recognized the cost effectiveness as well, resulting in subsequent casters built based upon the (SMS) Concast USA technology.

The Japanese requirements for beams started to increase dramatically in the 1980's with the great increase in their building and construction industry and being located in a severe earthquake zone. This increase escalated and today, their requirements exceed eight million tons annually, approximately two times greater than that of the structural demands in North America.

At this time in the United States there was no great demand for beams, due to minimal need for rebuilding the infrastructure. The supply of beams was for years produced primarily by the large integrated steel plants, specifically Bethlehem Steel, U. S. Steel and Inland Steel. The mini-mills became very successful in the 1970's producing bar products. Three of these mini-mills recognized that they also could be a producer of the smaller H & I beams and could produce these lightweight sections at a cost far less than that of the three integrated producers. This led Nucor Steel to form a joint venture with Yamato Steel of Japan, in 1988. The originally developed technology could now be successfully implemented at a rapid start-up rate.

The Nucor-Yamato Steel Blytheville, Arkansas structural steel producing facility consisted of an SMS Concast supplied three strand beam blank caster which produced wide flange beams up to 27 inches in width. The rolling mill that was used was basically conventional, however, labor costs were exceptionally low and the efficiency of the caster and mill was such that it provided a huge cost advantage over the integrated producers.

In 1986 SMS Concast modified the existing Northwestern Steel and Wire existing six-strand machines to produce beam blank shapes for small lightweight beams.

The utilization of tube moulds for these beam blank sizes was developed. At the same time, the Chaparral Steel five strand machines was also converted to use a tube beam blank mould for the casting of their small section. This section would feed their medium section mill and produce small lightweight beams.

The successful development of the casting process depends mainly on the understanding and application of the fundamental knowledge of transport phenomena. The heat transfer and stress have very important influences on product quality, such as: crack formation, product structure, property, and so on. For many years, a great deal of research on heat transfer and stress, in continuous casting, has been done, which provides a lot of useful information for practical production. These models, however, have been confined to continuous casting in simple geometry, such as, billet, bloom, and slab. In continuous casting of the beam blank, various kinds of defects can occur mainly because of its complex shape. As can be seen from its shape, temperature, and stress profile, the solidification patterns in the continuous casting of the beam blank are different from those in the casting of simple shapes. Unfortunately little information is available in the literature on the simulation of the near net shape casting, especially on thermal-mechanical analysis.

7.3 Thin Slab Casting

Over the last decade steel producers from the United States, Europe and Japan have invested considerable research into the development of direct-casting processes. The interest has stemmed from a desire to reduce production and investment costs by simplifying the overall steel-making process. The simplification results from the casting of a steel product close to or at the necessary thickness for cold rolling. The elimination of most if not all process steps between casting and cold rolling results in a much less complicated process, with energy savings and lower investment costs. Direct-casting research has focused primarily on three traditional process configurations; the twin-belt, twin-roll, and single-roll arrangements. In addition considerable research is being directed toward newer innovations such as spray forming and electromagnetic-levitation casting processes. Refinements to the twin-belt process have been for the purpose of casting thin slabs, approximately 25-75mm in thickness, whilst for double-and single-roll processes the emphasis is on thin strip sections, typically 1-6mm and 20-500mm respectively. Thin-slab product is typically not sufficiently thin for direct introduction into cold-rolling facilities and in general requires some additional hot rolling.

Thin slab casting can provide near-net-shape products, and it requires less rolling process, thus it is more economical compared with conventional slab casting. In the thin strip steel casting process, molten steel is retained between two rotating rolls and two fixed refractory side dams. The shape and size of the plates obtained are determined by the roll diameter and the depth of molten steel in the tundish. The fixed side dams sliding against the rolls are subjected to high stresses due to localized temperature differences. The plate temperature varies from steel melting temperatures of

1500-1600℃ to roll contact and back side dam temperatures of 200-400℃.

7.3.1 Problem

(1) Compared with conventional slab continuous casting, the casting speed of thin slab casting is more than 6m/min higher, and its mould has a large wide to narrow sides ratio, hence the fluid flow in the mould is very difficult to be prevented from the non-uniform initial thin solidification of shell, severe oscillation of meniscus, and some other defects in the casting process. Several numerical simulations of fluid flow, or fluid flow combined with other phenomenon. A detailed review and forecast can be found in the previously published studies. However, most of these studies are based on simple geometrical systems with rectangular or circular sections, thus relatively light and easy computational fluid dynamical studies are required. Honeyands and Herbertson have made pioneer fluid dynamics researches on thin slab casting mould using water model. To reduce the turbulence of feeding liquid steel and its impingement to the mould wall, the superstructure of thin slab casting mould is designed as funnel type. Correspondingly, SEN(submerged entry nozzle) is specially designed with prolate outlet section. Nam et al. performed numerical analysis of fluid flow coupled with heat transfer and solidification in a funnel type thin slab casting mould using finite volume method that is based on structured body fitted coordinate grids. Characteristics of the transport phenomena in the mould were numerically analyzed. Thereafter, Park et al. performed numerical analysis for parallel type thin slab mould, which is similar to the above study, and increased attention was paid to the optimum design of SEN, which has considerable influence on flow pattern, heat transfer, and accordingly the solidification process.

(2) Side dam wear is not the only problem to be considered in analyzing this process. Thermal shock and high stresses due to localized temperature differences; chemical interactions between the side dams and the molten steel; and the unexpected development of a partially solidified steel shell on the rolls may lead to rapid side dam damage. Maximal side dam tensile stresses are developed in the roll contact area. At the bottom part of the side dams, cracks may occur leading to molten steel infiltrations into the contact area and to an erosion of the side dams.

This erosion has to be compensated by a mechanical wearing of the refractory plates by the rolls to avoid and prevent such infiltrations by maintaining a close contact between the rolls and the side dam. The mechanical wearing was obtained by a constant displacement of side dams according to an experimental knowledge. This method led to a significant consumption of side dams and limited the performance of this process.

This mechanical wearing has to be controlled from the side dam bearing the load. This method can be applied if refractory wear depends on the contact pressure between side dam and rolls. Controls of the mechanical wear process require understanding the mechanisms of third body formation and the velocity accommodation mechanisms(VAM). Wear control also requires the identification of wear parameters governing the wear of SiAlON-BN and the quantification of their influence.

7.3.2 Practice

The use of aluminum in the automotive industry is nowadays a well-established practice, which enables manufacturers to improve vehicle fuel economy and reduce CO_2 emissions. The advantage of aluminum over competitive materials is its very attractive combination of low density, high strength and formability, ease of recycling, and high corrosion resistance. While the application of aluminum die-cast parts is extensive, the use of aluminum sheet is relatively restricted. The major barrier to the widespread use of aluminum sheet in automotive applications is its high cost, which is four to five times that of steel sheet. The production of aluminum sheet by twin-roll continuous(TRC) casting, rather than by conventional direct-chill(DC) casting and hot rolling, offers an opportunity to reduce sheet cost substantially.

Due to the high solidification rate achieved in TRC casting, the microstructure of TRC-cast materials differs significantly from that of DC-cast ones. TRC alloys usually exhibit high concentration of alloying elements in solid solution, fine primary inter-metallic particles and fine as-cast grain size. All these features affect alloy response to thermomechanical treatment involved in the downstream processing. Therefore, the microstructure of sheets issued from TRC-and DC-cast alloys can differ markedly. The differences in microstructure and in crystallographic texture have great impact on sheet mechanical properties and formability.

Aluminum sheets for automotive application should exhibit(besides other properties) good formability in combination with sufficient strength. The DC-cast materials currently used in such applications fulfill this requirement rather well. Recently, there has been an increased interest in using TRC as a method to produce low-cost/high-quality Al-Mg(AA5xxx series) sheets for automotive structural applications. The research works on TRC-cast Al-Mg alloys are concentrated on determining the parameters of downstream processing in order to obtain final gauge products with properties equivalent to, or better than, those of DC-cast alloys. Therefore, a detailed knowledge on the microstructure of the sheets produced by both casting methods, including their crystallographic texture, is necessary.

7.4 Strip Casting

The twin roll strip steel casting process, developed by the neither using nor company under a project called Myosotis, can be illustrated by Fig. 7-3 Twin-roll strip casting. Molten steel is introduced between two nickel-coated copper rolls. Two refractory side dams in contact with the rolls contain the molten steel. The rolls are water cooled, with a diameter of 1.5m and a length of 1m. The gap between the two rolls determines the thickness of the steel plate. At the beginning of the study, this new process was able to transform during one cast approximately 90t of molten steel into a plate 3.5km long, 0.865m wide and 2.8mm thick. The contact area of the side dam consists of a SiAlON-based BN composite, developed by the Vesuvius Company.

Twin-roll strip casting is regarded as a prospective technology of near net shape continuous casting. The thickness of strip produced is approximately 1-6mm. Although thin strip casting has ad-

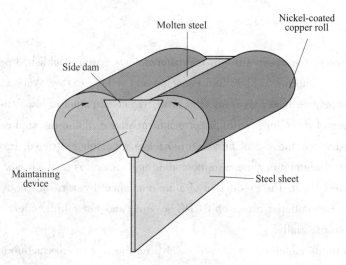

Fig. 7-3 Twin-roll strip casting

vantages of low energy consumption, low investment and reduced emission, it has some problems of smooth operation and product quality.

During strip casting, the flow field pattern and level fluctuation in the pool affect the strip quality. When the level fluctuation exceeds ±1.6-2mm, the longitudinal cracking will be observed. On the contrary, if the meniscus is too calm, the turbulence would be strengthened in lower part of the pool and the surface layer would solidify. Besides, the configuration of feeding system, casting speed and pool depth also affect flow field. So it is very difficult to control flow field in the pool. At present, the key factors influencing the strip quality are dam edge, feeding system and characteristics of rolls. The configuration of the feeding system is of decisive importance to flow field in the pool. The molten steel must be supplied homogeneously over the pool width, because asymmetric flow field pattern influences the solidification negatively. The flow field pattern is affected by the geometry of feeding system as well, and the heat transfer in the mould formed by two roll surfaces is affected by the flow field pattern.

In Several years, research has focused on questions like how to save energy, process time and money as well as how to get new materials with new properties. Concerning flat material, one of the answers given is thin-strip casting by double-roller technique. It is a complex process whose final product is a hot strip of 2-7mm thickness. Complexity is caused by combining solidification and forming in one single step. All requirements to the final product have to be adjusted during the few parts of a second of casting.

The casting machine at the IBF, which is operated together with Thyssen Krupp Stahl, consists of an induction furnace, a double roller caster with two horizontal rolls with parallel axis, a pinch roller and a coiler. A water spray cooling zone can be linked into the process chain if required. The melt is cast between the rolls via a submerged nozzle. Ceramic side plates seal the front sides. Solidification takes place at the rotating copper sleeves of the rolls as two strip shells. They merge at the kissing point in or just above the narrowest roll gap. A completely solidified strip leaves the roll gap.

7.4.1 Twin-roll casting

Despite the earliest patents relating to twin-roll casting as a method for producing metal strip directly from the melt dating back to the middle of the last century, it took over a hundred years for the process to be exploited commercially. In the early 1950's, Hunter Engineering introduced a twin-roll caster for the production of aluminum strip. In these early machines the water-cooled casting rolls were arranged horizontally and were fed from below by means of a refractory feeder tip. The roll diameter of the earliest machines was 600mm and the rolls were clamped together hydraulically with a force of approximately 600t. The strip produced from this type of machine was typically 6mm thick and of up to 1500mm wide.

After leaving the casting rolls the strip was allowed to form a loop before entering strip handling equipment comprising a pair of pinch rolls, a static shear, a four-roll bridle and finally a coiler. Although machines of this type were adopted widely for the production of aluminum foilstock materials, their use for other applications was limited by a number of process and equipment problems.

Since the introduction of the Hunter vertical casters nearly 40 years ago both the equipment and the casting techniques have evolved gradually, but there has been little or no change in the underlying process technology. Machines with larger rolls—typically 1000mm diameter and 2000mm wide—are being used currently and it is usual for the rolls to be arranged vertically or tilted slightly so that the cast sheet emerges horizontally from the rolls. It is possible to cast a range of alloys, but the vast majority of material produced by this method is pure grade (narrow freezing range) alloys for applications such as foilstock or finstock. In spite of the relative simplicity of the equipment and the lack of significant process development, the product from twin-roll casters is suitable for downstream processing, with many operators producing finished foil products at or below $7\mu m$ thickness.

7.4.1.1 The process

From an equipment standpoint, a twin-roll caster is essentially a slow, two-high rolling mill where the feedstock is molten aluminum and the end product is cast aluminum strip of 6~10mm thickness. Solidification begins when the molten metal contacts the water-cooled casting rolls, and due to the progressively reducing dimension of the roll bite, the solidifying metal is forced to remain in contact with the rolls. Once solidification is complete, the material undergoes a degree of hot working before leaving the roll bite.

Depending on the operational parameters, significant separating forces can be developed during twin-roll casting and it is because of these high loads that twin-roll cast sheet is characterized by good surface quality and good gauge and profile tolerances. Even for relatively high purity materials such as alloy AA 1145, unit loadings of 0.5t/mm of width are the norm. For long freezing-range alloys such as AA 5182, loads approaching 1t/mm of width are common. As a result of these high loads significant deformation of the casting rolls is observed. By way of example, the loads generated when casting 1500mm wide alloy AA 1145 are sufficient to deflect the caster rolls by up to 2mm. A consequence of this level of deflection in the Easter rolls is that in order to produce nominal 6mm strip, the initial roll gap at the edge of the rolls has to be set at approximately 4.5mm. When the effect of roll camber is added to this-typically 0.75mm on the diameter of each roll-the roll gap at the centre line of the rolls is often about 3 ram, which can cause significant operational problems.

7.4.1.2 Equipment evolvement

One example of a modification to machine design that has been of significant benefit to caster users, and has improved the quality of the cast product, is the method of setting and adjusting the roll gap during casting. On the earliest machines the roll gap was set by inserting packers between the chocks, and the gap measured with feeler gauges. Setting the gap was a laborious process and once the cast was underway there was no way of adjusting the gap. Side-to-side thickness variations in the strip could be overcome only by adjusting the tip set-back on one or other side of the caster, but this can cause other problems. On later machines the simple mechanical packers were replaced by an adjustable wedge system. With this system it is possible, at least in principle, to adjust the roll gap during casting, but this procedure requires that the hydraulic clamping force is reduced essentially to zero to allow the chock separation to be altered. By its very nature this procedure is a "hit or miss" affair and several attempts are needed to correct any sheet deficiencies. A major advance was made by Davy in the mid-eighties when load cells were used in conjunction with wedges.

The separating force the cast sheet generates is a function of the strip thickness and the major operating parameters such as casting speed and tip set-back. Provided the separating force can be measured accurately, it is possible not only to control the side-to-side thickness of the strip but also the profile of the cast strip. The gauge and profile-tolerances of continuously cast sheet have become increasingly more demanding and this has brought about a fundamental change in the attitude of machine builders and caster operators alike. It is quite usual for the side-to-side gauge variation in the cast strip to be within 1% and for the strip profile to be between zero and 1% positive.

The applied clamping force is opposed by the separating force generated in the cast strip, the resultant force-the chock force-being measured using load cells positioned between the chocks. To produce parallel strip the chock force on each side of the machine needs to be the same, and to produce the required profile the total separating force needs to be within pre-defined limits.

As the separating force increases from zero to its steady-state value, the shape and dimensions of the roll gap change and the sheet profile goes from being negative, through parallel, to being positive. The required profile is between parallel and 1% positive and to maintain this profile it is essential that the separating force is controlled between closely-defined limits.

The instantaneous value of separating force is measured using load cells, and can be used as the measured variable of a control system. Three loops, caster-control systems based on the continuous measurement of separating force are in operation on a number of commercial casters. Even though this development is a major advance on earlier systems, changing the roll gap during casting remains difficult and is only normally attempted when all other avenues have been exhausted. The ultimate solution is to replace the wedge assemblies completely and to set the caster roll gap hydraulically using position regulated cylinders. Using this technique, changes in roll gap can be achieved effortlessly simply by changing the reference signal to the hydraulic cylinder. The caster control strategy remains essentially the same, the only change being that instead of measuring the separating force with load cells between the chocks, it is measured by the use of pressure transducers located in the hydraulic cylinders. The roll gap is measured either by transducers located between the chocks or rolls or by transducers located in the hydraulic cylinders.

7.4.1.3 Experimental program

Since the installation of an experimental twin-roll caster at the University of Oxford in early 1989 as part of a SERC/DTI-supported Teaching Company Scheme, a series of over 600 casts have been carried out. As originally constructed, the caster comprised a pair of internally water cooled steel rolls of 400mm diameter and 300mm face width arranged so that the cast strip emerged horizontally, only the bottom roll was driven and the roll gap was set manually using a series of packers before each cast. Details of the equipment and the development programme are reported elsewhere.

Numerical modeling of the process has predicted increasing productivity with decreasing strip thickness, and the initial casting trials were designed to establish the operational capabilities of the caster and to examine the validity of the modeling results. As has been mentioned earlier, roll deflection and ground-in camber make the startup of wide casters difficult and this problem is exaggerated at thin gauges. This was not a serious problem with the narrow width experimental caster due to the inherent stiffness of the caster rolls, although as a result of only driving the bottom roll, motion transfer to the non-driven roll was too slow to enable start up at thin gauges without premature freezing of metal in the roll bite.

To overcome this difficulty the experimental caster was modified to incorporate hydraulic gap control, independent drive to each roll and a lever actuated tip table. With this arrangement it was possible to start the caster at a nominal 4mm gauge and reduce the roll gap progressively, during the cast. By making corresponding increases to the caster speed, it was possible to produce strip down to 1mm gauge at speeds of up to 15m/min. The lever-actuated tip table was incorporated to facilitate retraction of the tip as the roll gap is reduced whilst maintaining an effective seal between the tip and both rolls.

The success of thin-strip casting is governed by the control of numerous process variables. For each alloy and gauge it has been found that successful casting can be achieved only between set limits-the operational window. Factors that affect this window are, inter-alia, casting speed, tip setback, metallostatic head, melt superheat, grain refinement, applied strip tension, roll lubrication, roll cooling and roll surface condition. The size of the operating window for each alloy is governed by the formation of defects such as heat-lines, centerline segregates, sticking and hot tearing, as well as the mechanical and micro-structural properties of the strip. The cast-ability of an alloy can be measured in terms of the relative size of this window. Generally, it has been found that cast-ability decreases with decreasing thickness as the process becomes more susceptible to small changes in process variables, and for many alloys the process is of limited stability at thin gauges. As a result, manual operation or even simple automatic control of the process becomes increasingly difficult, and it was necessary to develop a control system capable of interpreting the condition of the casting process and controlling the process variables to maintain strip gauge, quality and productivity at their optimum levels.

A wide range of alloys have been investigated using the experimental caster. Although predominantly aluminum based, the range of alloys is extensive, varying from ultra-high purity material to wide-freezing-range alloys of Al-Sn, Al-Cu, and composites.

As a result of these investigations it has been found that, in general, the operating window for wide-freezing range alloys is significantly greater than for narrow-freezing-range alloys, which makes them particularly useful for understanding the casting process at thin gauges and high speeds. A detailed study of the effect of process variables on the casting of AA 5182 has shown how the casting process is influenced by the choice of casting conditions, and how increases in productivity can be achieved by casting at thinner gauges, as predicted by mathematical modeling. As a result of this understanding it has also proven possible to cast both narrow and intermediate-freezing-range alloys at gauges down to 1mm and speeds of up to 15m/min. These limits are imposed by the mechanical design of the caster rather than by process problems, and further improvements in productivity are envisaged when the current round of machine modifications are completed.

7.4.1.4 Sticking

During twin-roll casting it is quite common for the strip to stick to one or both casting rolls.

If sticking is light, casting can continue, although due to the increase in strip thickness that results from the improved heat transfer during sticking, the material may have to be scrapped. In cases of severe sticking the cast usually has to be terminated, as the strip bonds tenaciously to the roll and cannot be removed while casting. Removal of 'stuck' strip from the rolls invariably causes damage to the roll surface that in extreme cases necessitates the regrinding of the rolls. To minimize this problem it is standard practice to spray the roll surfaces continuously with lubricant. A wide variety of materials have been tried but the most widely-used lubricant is a suspension of colloidal graphite in water. It is normal to apply the lubricant by means of air atomizing nozzles and these can either reciprocate or traverse the caster rolls. Sticking can still occur even if lubricant is sprayed onto the caster rolls and in this event it is normal practice to slow down the caster until the problem becomes manageable.

From the work carried out on the experimental machine at Oxford, it is apparent that the phenomenon of sticking is a major impediment to high-speed, thin-strip casting. Moreover, the character of sticking observed at high speeds and thin gauges is different to that observed at conventional speeds and gauges.

These observations can be summarized as follows:

At strip thicknesses in the range of 4-6mm, sticking occurs at high speeds; i. e. , "hot" casting conditions. The rate of application of lubricant is increased and/or the casting speed is decreased to give the maximum strip speed without sticking. If sticking does not occur, heat lines, which are another characteristic defect associated with twin-roll casting, are likely to develop. Whilst there are no hard-and-fast rules regarding which alloys stick, in conventional twin-roll casting it is a general observation that the tendency to stick increases with increasing alloy content.

Sticking is more noticeable as the thickness of the cast strip is reduced to 1-2mm. Alloys with freezing ranges greater than approximately 50°C do not stick even at high casting speeds; e. g. , 15m/min. Alloys with an intermediate freezing range tend to stick when the speed is reduced, whereas narrow-freezing-range alloys tend to stick when the casting speed is increased.

When sticking occurs, the forward slip disappears, forward slip being the difference between the strip speed and the peripheral speed of the casting rolls, which for normal casting conditions has been found to be typically 5%-10%.

The roll temperatures increase at the onset of sticking, and there is a corresponding decrease in strip temperature and increase in strip thickness(and separating force).

Too much lubricant leads to poor surface quality and the possibility of other casting defects, whereas too little can promote sticking. The rate of application of lubricant is critical and must approximate closely to the rate it is removed.

To overcome the problems of sticking, particularly when casting at high speeds and thin gauges, a number of solutions have been proposed and evaluated.

(1) Mechanical methods.

Whereas existing spray arrangements are adequate for conventional casting-narrow freezing range alloys at low speed and thick gauges-they have been demonstrated to be of limited use under high-speed, thin-strip casting conditions. The main reason why they are inadequate is that the coating is not uniform across the full strip width, this being the result of the method of application.

The parting agent is usually applied to the casting rolls in the form of an atomized spray from either traversing or reciprocating spray nozzles. Neither of the aforementioned systems deposits a uniform coating onto the caster rolls. With both systems the lubricant is deposited in a helical pattern, the pitch of the helix being a function of the speed of rotation of the rolls and the traverse rate. As a consequence, some areas of the roll remain uncoated whereas other areas, particularly near to the end of the traverse, receive a double layer of coating. The obvious lack of uniformity is deleterious, as the thickness of the parting layer affects the rate at which heat is removed from the cast strip. In areas that remain uncoated there is a tendency for sticking to occur, whereas in areas where there is an excess of lubricant the rate of heat removal is diminished and other casting defects such as heat lines can result. To overcome this problem an improved method of applying lubricant has been developed.

It is a common observation that when sticking occurs the forward slip decreases from approximately 10% to zero. This means that whereas during normal steady state "non-sticking" casting the strip speed is 10% greater than the peripheral roll speed, once sticking occurs the strip speed is forced to travel at the roll speed. As a result of this observation it has been proposed that if the strip speed could always possess a scrubbing motion relative to the casting rolls, then sticking would not occur. This effect can be achieved by modifying the arrangement of the roll stack so that instead of the axes of the casting rolls remaining parallel they are crossed relative to each other. By crossing the rolls a slip component is generated at right angles to the casting direction and it can be shown that only very small amounts of crossing are necessary to generate the amounts of relative motion normally associated with "non-stick" casting. Roll crossing is an established technology in rolling mills

Where it is used for the control of strip profile and, although it adds complexity into the mechanical design of the machine, the potential benefits make it worthwhile.

(2) Chemical methods.

For high-speed, thin-strip casting, the presence of residual liquid on the surface of the cast strip during the latter stages of solidification has been shown to be beneficial in avoiding sticking due to the self-lubricating effect. This effect occurs naturally in wide freezing range alloys but does not occur in the majority of alloys that are produced using the twin-roll technique. Moreover the composition of these alloys has to be maintained within closely defined limits to meet internationally agreed standards. Consequently the problem can be summarized as follows; "The effective freezing range of the alloy has to be increased dramatically without putting the alloy out of specification". When one considers that, although the precise figures will depend on the alloy system in question, significant changes in composition are required to bring about even modest changes in freezing range, the task appears impossible unless some 'magic' ingredient can be found.

Examination of the aluminum-alloy phase diagrams confirms that such alloying elements exist and experimental evaluation of candidate materials has demonstrated that dramatic improvements in sticking behavior can be achieved by minute additions of particular alloying elements.

(3) Process solutions.

For the case of strip 4-6mm thick, the conventional method of overcoming the problem of sticking is to progressively reduce the casting speed until sticking ceases to be a problem. By slowing the machine down (or alternatively increasing the tip set-back) the separating force is increased, this having the effect of increasing the degree of forward slip. The increase in relative motion between the strip and the rolls is discouraging sticking. By running the development caster with only one roll driven it was possible to establish the role of forward slip in the prevention of sticking.

When only one roll is driven the other roll is driven by the cast strip. This means that the strip travels at essentially the same speed as the driven roll (zero forward slip) whereas the non-driven roll travels slower than the strip speed; i.e. the strip slips relative to the non-driven roll. Operating under these conditions it is noticeable that sticking never occurs on the non driven roll and that, whereas the appearance of the driven roll remains unchanged during casting, the non-driven roll develops a lustrous surface coating. From these observations it is apparent that the relative speeds of the rolls are of significant importance and this is currently being investigated.

Although twin-roll casting is an established technique, the process has limitations both in terms of product quality and productivity and is only used for a limited range of products. Modifications to the design of the equipment have improved the operation and control of the process and brought about significant improvement in the gauge and profile tolerances of the cast strip. As a result of a four-year experimental program, during which over 600 casts using a wide range of materials have been carried out, the level of understanding of the process has been expanded and this has led to dramatic increases in productivity and the ability to cast alloys that hitherto were regarded as "uncastable". Further development is underway, which will include proving the concept of high-speed, thing-strip casting at commercial widths to this end a Joint Developmental agreement.

7.4.2　Thin strip casting of high-speed steels

The strip casting process, which produces strip sheets directly from molten metal, is drawing much interest as a prospective technique in the steel industry, the reason being that the development of the strip casting process may realize not only the shortening of processes and the saving of energy consumption, but also the near-net shaping of difficult-to-form materials. One of the most promising approaches amongst various methods for thin strip casting is the twin-roll method. A lot of experimental and analytical studies have been carried out on this method in the past decade. Most of such work being focused on the strip casting of stainless steels and silicon steels. At present, pilot tests for the stainless steels have been carried out in at least five countries. Little work, however, has been done on the thin strip casting of high-speed steels.

High speed steels are very important tool materials, their use including the manufacture of hacksaw blades. The thickness of the strip sheets for the hacksaw blades is normally less than 2mm. When the conventional technique is used to produce the steel sheets, multi-rolling and multi-annealing is unavoidable in order to decrease the thickness of the products and to break up the as-cast carbides. As a result, the production Process takes a long time. In this investigation, the twin-roll method is used to produce the strip sheets in order to shorten substantially the production process, the metallurgical quality of the strips the microstructure, focusing on the carbides, as well as the cutting performance of the hacksaw blades, being examined.

连铸工艺与技术

1 引 言

1.1 连铸工艺简介

连铸工艺的基本原理为将钢液浇入无底的水冷铜结晶器。现代连铸机的简单描述有助于读者领会连铸工艺的各个方面。图 1-1 为现代板坯连铸机总体布局示意图，钢包位于钢包回转台上，钢包回转台可以旋转，因此盛满钢液的钢包可以迅速进入连铸位置，保障浇注的连续性。

钢液最初从炼钢炉注入钢包，经过适当的二次精炼后，由钢包运载装置吊运至连铸机上方，钢液通过水口从钢包注入中间包，钢液注流由耐火水口保护，以避免空气的二次氧化。中间包和结晶器之间的钢液注流也是由耐火水口保护以避免空气氧化。铸坯沿铸机方向行进几米后完全凝固，然后经矫直辊矫直。在矫直点，铸坯水平运行并通过动力驱动辊送出铸机。

1.2 连铸的发展及机型演变

相对于氧气炼钢工艺，钢的连铸（CC）工业化的历史相对较短。在萌芽阶段，即 20 世纪 50 年代前的相当长的时间，连铸工艺的发展源于有色金属行业。有色金属已经采用轮式连铸机或带式连铸机浇注，特别是采用运动式结晶器，克服连铸过程的摩擦力。虽然当时的连铸工艺主要基于固定的振动式结晶器，钢铁从业者的真知灼见对推动连铸工艺在钢铁工业的应用有里程碑式的意义。对于世界钢铁工业而言，1970 年连铸坯占粗钢产量只有 4%，2018 年已达到 96.2%。

连铸的发展十分迅速，沃尔夫博士把连续铸钢的发展概括为 40 年代的试验探索、50 年代开始步入工业化、60 年代弧形铸机出现引发一场革命、70 年代两次能源危机推动大发展、80 年代技术日趋成熟和 90 年代以来面临高生产率、高速浇注、高质量产品、节能环保、改善劳动者工作条件、过程和质量控制技术和近终形连铸（薄板坯、薄带连铸等）等的不断挑战和创新并取得显著成就的 6 个阶段。尽管今天的连铸已经达到相当成熟的高度，但以振动结晶器为标志的常规连铸法存在的 3 大问题，即表面振痕、凝固组织和传热机制，依然是有待克服的技术障碍。

连铸机机型的演变是一件有趣的事。早期的连铸机是完全垂直的，但是这种连铸机需要相当大的高度来实现每铸流的生产效率，并且随着产能超过 400t/h 的碱性氧气转炉炼钢法的快速发展，连铸机需要更多的铸流数以实现浇注速度与炼钢炉产率配合。过去 30 年来不断发展的连铸机设计如图 1-2 所示，出现了从立式连铸机（铸机 1）到"低头"连铸机（铸机 5）很多机型。

1965 年，连铸机的构造还非常简单。80% 的连铸机是立式连铸机，用于板坯、大方坯和小方坯连铸。之后弧形连铸机占主导，在 1975 年，80% 的板坯连铸机和 70% 的大方坯和小方坯连铸机是弧形的。连铸机型的发展趋向更复杂的几何形状，采用渐进弯曲和矫直，并于 1984 年应用于 30% 的板坯连铸机和 20% 的大方坯和方坯连铸机。

(1) 立式连铸机。垂直浇注具有其天然的技术优势。由于铸坯是从结晶器中脉冲拉出，铸坯冷却均匀确保了铸坯组织均匀和可预测的晶体生长模式以及凝固坯壳上的轴向载荷均匀分布。在管材或空心型材连铸中，该工艺具有特殊优势，因为它去除了末端氧枪，可保证铸机末端位置的产品质量。在铸造贵金属时，后者的优势非常重要。

(2) 立弯式连铸机。立弯式连铸机铸坯导向段由以下部分组成：1) 与垂直段相连的弯曲段；2) 与弯曲段相连的过渡矫直段；3) 与过渡矫直段相连的主导向段；4) 设置在主导向段和水平段之间的最终过渡矫直区。

弯曲段具有不止一个相对较强的顶弯力的初弯区，顶弯力被一系列从垂直段的无限大曲率半径开始逐渐减小的曲率限定，导向段和二弯区与初弯区相比，顶弯相对平滑。在顶弯段和主导向段之间的过渡矫直段被不断增加的铸坯曲率半径限定，并且顶弯段的第一和第二区以及过渡矫直区均呈椭圆形。

在立弯式连铸机中，直结晶器和二冷区直线段不仅可以均匀冷却钢液和带液芯的铸坯，也是保证结晶器和二次冷却段具有均匀热通量的基本条件，更是生产优质铸坯的重要基础。立弯式连铸机铸坯的重力可以代替拉坯装置的拉力，改变受力状态并减小板坯所受的外力（拉力），最终改善板坯质量（见图 1-3）。

(3) 弧形连铸机。立式连铸机利用重力浇注并确保铸坯具有均匀的宏观组织，是最原始的机型。但是连铸机的产能受到铸机高度的严重限制。因此，连铸历史上的一些通过铸坯弯曲和矫直来延长在低厂房前提下的铸机长度，例如 Rowley 研发的方坯连铸机和 Tarquinee 和 Scovill 提出的进一步将温度均匀的铸坯的在线尺寸考虑在内的提议，这一设计理念后来由美国钢铁公司为南方企业提供试运行铸机，并且在加里企业 1 号板坯连铸机上首次得以实现（见图 1-4）。弧形连铸机的半径大多在 9~12m，中冶京诚研发的连铸机可以浇注直径 1000mm 的圆坯，铸机半径达 17m，铸机长 37.6m。

在弧形连铸机中，铸坯垂直（或近乎垂直的弯曲）地离开结晶器，通过喷淋室时，支撑辊逐渐引导铸坯，制止其水平运行。减少连铸系统整体高度，首先需要在钢液进入直结晶器后并在弯曲之前完全凝固，或者铸坯带液相进入到弧形结晶器，这是最常用的方案。垂直连铸系统和铸坯完全凝固后被顶弯的系统具有长且直的液相，均导致资金投入过高。然而，从维护成本的角度来看，这些系统具有优势。在铸坯仍含有液相的情况下顶弯的垂直系统的优点是，厂房不需要与凝固后被弯曲的铸坯一样高；然而，液相弯曲系统兼顾了低的投资费用和维护成本。

为了防止内部裂纹，基于固/液界面处的临界应变和应变率的连铸机机型设计的若干准则得以发展，这导致弯曲和矫直段的长度延长了多个辊对。

（4）水平连铸机。水平连铸机结晶器轴线是水平的，并且从液态到薄坯壳再到完全凝固（无弯曲）的运行轨迹是水平的。此类连铸机中，铸坯振动或结晶器振动以防止坯壳与结晶器壁黏结。这样做的原因主要是在铸坯处理难度低的基础上，便于操作和保证安全。当然，应用水平而不是垂直铸造的原因主要涉及由重力引发的定向冷却的难题；不过在大多数情况下，可以克服这些困难（见图1-5）。

1.3 连铸的优势

与模铸相比，连铸具有相当大的优势。模铸包括至少一个额外的加热和轧制过程以生产基本相似的铸坯，而这些铸坯在连铸工艺中以小方坯、大方坯或板坯的形式直接生产。

连续铸造工艺相较于上述模铸工艺的主要优点如下：提高金属收得率，减少能源消耗，节省人力，提高产品质量，降低对环境和工人有害的气体排放，降低库存、缩短交货时间，减少新钢铁厂的投资成本。

相较于模铸，连铸工艺能源消耗低的原因为：省去重新加热段；利用钢液带来的热量，在提高金属收得率的同时节省了能源。

模铸的工艺路线需要铸坯既要在铸锭模成型，又要在轧制成半成品即板坯、大方坯或小方坯后进行加热。在连铸流程中，第一个再加热阶段被铸坯轧制前的预热所代替。

2 连铸设备

2.1 连铸机主要参数

2.1.1 连铸机机数和流数的定义

连铸设备是保障连铸良好运行的最基本、最重要的因素。一套连铸设备具有独立的动力系统和独立的工作系统,当其他机器出现故障时可以继续工作。一套连铸设备可以自成系统,也可以多台合作。一套连铸设备可以同时浇注一流或多流铸坯。本身具有拉坯装置和驱动装置的机器被称为连铸机。一台连铸机可以有一个或多个拉坯装置和驱动装置,并形成一个或多个铸流,因此有1机1流、1机2流、2机2流等类型的铸机。

定义:(1)台数。凡是共用一个钢包,浇注1流或多流铸坯的1套连铸钢设备称为1台连铸机。(2)机数。凡具有独立传动系统和独立工作系统,当其他铸机出现故障时仍能照常工作的一组连续铸钢设备,称之为1个机组。1台连铸机可以由1个机组或多个机组组成。(3)流数。1台连铸机能够同时浇注铸坯的总根数称为连铸机的流数。1台连铸机有1个机组,又只能浇注1根铸坯,称为1机1流;若1台连铸机有多个机组,又同时能够浇注多根铸坯,称其为多机多流;1个机组能够同时浇注2根铸坯的称为1机2流。

2.1.2 铸坯断面尺寸

表2-1是各种常见铸坯断面。根据铸坯的断面形状和尺寸,分为小方坯、大方坯、圆坯、板坯和异型坯。在某些情况下,特定铸机上的结晶器可以改变为其他断面形状;例如,小方坯改变为大方坯,大方坯改变为小板坯,大方坯改变为圆坯。此外,铸机可以浇注特殊断面形状,例如矩形坯和犬骨状异型坯。

除非严格控制二次冷却,否则含碳0.18%~0.28%的低合金钢难以浇注。然而,许多企业都浇注了很多的钢种,如SAE 8620。几家企业正在尝试连铸硼钢。不锈钢中厚板(奥氏体不锈钢)、不锈钢薄板(奥氏体不锈钢、铁素体不锈钢、马氏体不锈钢)、电工硅钢板、淬火和回火钢、表面硬化钢、弹簧钢和耐候钢也有巨大的应用潜力。

2.1.3 拉坯速度

拉坯速度是连铸机的一个铸流在单位时间(min)内浇注的铸坯长度(m)。显然,提高拉坯速度将提高连铸机的生产效率。因此,拉坯速度是连铸机非常重要的参数,会影响钢液的凝固速度和铸坯的内部质量。

假设板坯连铸机的凝固系数$K=27$或者大方坯/小方坯连铸机的凝固系数为30,并且采用的冶金长度L为15m、20m、25m,那么拉坯速度"v"和板坯厚度"d"相关,它们之间的关系由下式确定：

$$v = \frac{K^2 L}{(d/2)^2} \tag{2-1}$$

该公式在生产中具有很好的适用性；工作拉速和理论拉速遵循类似的关系，两者数值几乎相等。由于"K"的变化，钢种会影响拉速，例如不锈钢薄板坯的拉速通常在 0.8~1.0m/min 之间。对于大方坯/小方坯兼用型连铸机而言，工作拉速与理论拉速的偏差比板坯连铸机小得多，拉速的极值范围成了一个影响因素。例外是美国钢铁公司，因为质量要求很高，无缝管坯以低速（1.5m/min）浇注。

生产率是铸坯横截面的函数，单位是吨/分钟/流。板坯连铸机的生产效率要大得多，因为坯厚相同而板坯的横截面积更大。这些数据由统计的最小拉速、平均拉速和最大拉速计算得出。

2.1.4 弧形半径

弧形半径是铸坯的外弧半径，单位是 m。这是一个重要的参数，它决定了连铸机的高度和铸坯的厚度。

铸机半径可以通过经验公式确定，该半径是连铸机的最小弧形半径。

$$R \geqslant cD \tag{2-2}$$

式中，R 为连铸机的弧形半径，m；D 为铸坯的厚度，m；c 为系数。R 的取值范围通常是：小方坯连铸机为 30~40m，大方坯连铸机为 30~50m，板坯连铸机为 40~50m。

弧形半径也可以通过理论计算得出。

要求铸坯在进入拉矫机之前必须完全凝固，目的是不采用多对拉矫辊。冷却段的长度可以通过下式计算：

$$L_c = \frac{2\pi R\alpha}{360} + h \tag{2-3}$$

式中，R 为铸坯的圆弧半径；α 为圆弧半径中心所处的水平线和第一对轧辊轴线的夹角；h 为圆弧半径中心水平线和结晶器液面之间的距离。当我们采用弧形结晶器时：

$$h = \frac{L}{2} - 0.1 \tag{2-4}$$

式中，L 为结晶器的长度。当我们采用立式直结晶器时；h 为二冷区中直线段长度。

连铸机的圆弧半径必须保证冷却段的长度等于或大于液芯的长度。因此，连铸机的圆弧半径可以由下式确定：

$$R \geqslant \left(\frac{D^2}{4K^2}v - h\right) \times \frac{57.3}{\alpha} \tag{2-5}$$

式中，D 为铸坯的厚度，mm；v 为拉坯速度，m/min；K 为综合凝固系数；α 为圆弧半径中心水平线和第一对轧辊轴线的夹角。

2.1.5 冶金长度

从结晶器液面到拉矫机最后一个矫直辊轴线的实际长度称为冶金长度。它表示铸坯液芯长度的最大值。因此，连铸机的冶金长度与铸坯液芯长度密切相关。铸坯液芯的长度为：从结晶器钢液面到钢液完全凝固的距离。液芯长度可以通过凝固定律计算，铸坯完全凝固的厚度为 S：

$$S = \frac{D}{2} = K\sqrt{L} = K\sqrt{\frac{L_1}{v}} \tag{2-6}$$

整理后，可以得出液体核心长度 L_1：

$$L_1 = \frac{D^2 v}{4K^2} \tag{2-7}$$

式中，L_1 为液芯长度，m；D 为铸坯厚度，mm；v 为拉坯速度，m/min；K 为综合凝固系数，$mm/min^{1/2}$。

在设计铸机时，不仅要考虑铸坯可以达到的拉速，还要考虑连铸机投产后连铸技术的进步，以及可能实现的更快的拉速。因此，连铸机的冶金长度可以通过铸坯的最大厚度和最大拉速确定。

由于实际拉速始终低于最大设计拉速，因此液芯长度始终低于冶金长度。另外，连铸时拉速经常变化，液芯长度也随之变化。同一架连铸机，若拉速更快，则液芯更长，效益更好。

2.2 连铸机关键部件

2.2.1 钢包回转台

为保障连铸工艺不间断，需要快速更换钢包，钢包回转台是完成该任务的首选设备，可实现最快的钢包更换速度。钢包在大约 1min 内更换，相应地，中间包总钢流中断时间约为 3min。对于复杂的高产率连铸机而言，这是最常见的钢包更换方式。最简单的情况是，钢包回转台将钢包从固定高度的接收位置旋转 180° 至连铸位置。

钢包回转台的变体：钢包回转台有很多种变体。固定臂或升降臂钢包回转台以及独立或共同臂摆动式钢包回转台均有所应用。蝶型回转台是最受欢迎的高效钢包回转台。

蝶型回转台在每对臂上处理一个钢包，钢包总是彼此相对旋转，但能够独立升降，以便在连铸位提供中间包上方钢包的合理空间。这种回转台适用于所有尺寸的钢包。"C"型臂钢包回转台，用来应对钢包必须与连铸方向以 90° 角运动的特殊情况。

2.2.2 中间包和中间包车

2.2.2.1 中间包

中间包作为储存器可以容纳足够多的钢液，从而为结晶器提供连续的钢流，即使在更换钢包时也能如此，更替的钢包从炼钢流程按时供应。中间包水口通常为陶瓷水口，中间包出口位置与下方的结晶器位置相匹配。

中间包还可以作精炼容器，促使有害夹杂物上浮到渣层。中间包中的钢液表面通过添加粉末状覆盖剂，覆盖剂熔化后在钢液表面形成一层阻挡层，防止钢液氧化和热量损失。固体夹杂物颗粒如果留在铸坯中，在随后的轧制中可能形成诸如"条片缺陷"之类的表面缺陷，也可能引起局部内应力集中，降低钢材疲劳寿命。

中间包的具体作用为：(1) 作为允许更换钢包以进行连续浇注的储存器；(2) 作为杂质上浮容器，钢液在进入结晶器之前去除夹杂物；(3) 将钢液分配至各流。

为了满足这些要求，通常需要遵循的设计原则为（见图2-1）：(1) 钢液的平均停留时间约为5~10min，以促进夹杂物的上浮和去除，同时不会导致过度的温度损失；(2) 最小钢液高度为中间包出口上方2ft（600mm），以避免涡流，涡流可能导致钢液卷渣；(3) 在中间包出钢到最小钢液深度之前，钢包能够正常更换，同时连铸维持正常拉速；(4) 消除低流量或零流量的死区体积。

中间包入口处有一段抑制涡流的部分，其次是一段用于促进钢液流上升的部分。钢液表面被液态渣层覆盖，该液态渣层吸收由上升的钢液流带到液面的夹杂物并使钢液绝热，防止过多的热量损失。另外，在加入覆盖剂之前的钢液填充时间，或者中间包盖完全密闭的前提下，中间包表面可以吹氩保护。中间包外形、尺寸和流量控制部件（例如坝、堰和挡渣墙）的布局可使用物理水模型或理论流体动力学（CFD）模型来优化。

图2-2展示的是一个模拟实例，其中CFD已用于模拟整个中间包的温度，用于一系列流量控制。类似的模型也可以用于估计钢的温度变化，例如在小方坯连铸机的中间包中，将钢液分配到六个或更多单独的铸流，或用于预测钢液在中间包的平均停留时间、死区体积和夹杂物上浮情况。在长连铸时间的情况下，中间包钢液难以在连铸期间保持预期的过热度。为了缓解这个问题，可以应用中间包加热技术（见图2-3）。等离子加热系统适用于特殊情形，需要独立的支持系统，包括电源、水冷系统和火焰切割机的机械臂。

对于兼用型连铸，具体设计从中间包开始。中间包壳中有多个出钢口，并且根据铸坯的排列选择用于特定铸坯的出钢口。那些未使用的出口用耐火衬堵住。最简单的情形是，中间包位于连铸位置的固定支架上。耐火材料寿命通常为浇注10~20h。如果现代连铸从业者希望连续铸钢的时间更长，则必须在连铸期间更换中间包，这被称为中间包快速更换。该操作需要快速更换中间包，使相应注流的停浇时间最小。通常使用纵向行驶的中间包车实现这一目标。

中间包车的主要优点是，在配有两个中间包车的板坯连铸机上，即使其中一辆由于损坏或维护而离线，它们也可使连铸不间断。中间包车通常进行长距离和跨行程运动以及升降运动。这些运动可以是液压也可以是电机驱动。中间包耐火水口与结晶器的对中至关重要。

当中间包行至结晶器上时，中间包位置可能需要小的调整，以便水口在结晶器两宽面之间对中。在中间包车上，这种交叉行程通常由简单的手动液压泵完成。为了多炉连浇和减少快速更换中间包的次数，在连铸期间更换水口的需求不断增长。因此可以利用水口更换装置，可以在几秒内更换水口，同时对连铸操作的中断最小。

除了减少连铸期间所需的中间包之外，水口更换装置不仅减少了中间包渣壳，从而增加连铸机产量，而且使铸机具备紧急停机能力。

2.2.2.2 中间包车

除龙门式中间包车外，还有半龙门式和悬臂式中间包车，为结晶器操作工提供了更好的操作环境，但需要连铸平台上方的结构来支撑车轮。悬臂式中间包车从连铸平台水平离开钢轨，若敞开浇注，这种结构是有益的。但这种结构使得高架起重机的荷载波动不明确，例如用于设备更换时，需要更多的维护处理。悬臂式中间包车通常与小方坯连铸机上使用的较轻的中间包一起使用。

中间包快速更换通常可以在大约3min内完成，实际的注流（拉速）停止约90s。中间包车的替代选择是中间包回转台，原理上与用于钢包处理的钢包回转台类似。这种布局是紧凑的，但可能导致靠近结晶器的区域非常拥挤。因此，中间包回转台最常用于单流板坯连铸机。这适用于设备的尺寸不需要非常大，或者连铸平台空间有限的情形。

2.2.3 结晶器

结晶器是连铸机中唯一直接接触钢液的部件，是连铸机最重要的组成部分，必须能在恶劣的条件下运行。它需要通过有效均匀的传热来形成均匀的坯壳。结晶器还需要持久耐用，能够快速改变截面尺寸，并且仅需要最少的维护处理。

连铸结晶器被优质软化水冷却，冷却水由再循环系统供应。结晶器设计和故障安全系统通常使冷却通道中最小水流速度为8m/s。结晶器内部呈锥形以适应铸坯的收缩，锥度的大小取决于铸坯截面尺寸和拉速。

图2-4显示的是小方坯（a）、大方坯（b）和板坯结晶器（c）的基本结构。铜结晶器由钢衬板支撑，在结晶器的底部和顶部分别有进水管和出水管。在板坯和大方坯结晶器铜板背面，冷却水槽从上到下机加工而成，约15mm深、5mm宽。在小断面结晶器中，冷却水槽通常是管式铜结晶器和衬板之间的平行间隙。为了确保铜壁表面存在薄边界层且不发生沸腾，在这些冷却水槽中水流需要达到高雷诺数，这需要水速大于8m/s。

以下是两种主要的结晶器类型：（1）管式结晶器。通常用于铸造诸如小方坯等小断面铸坯。结晶器铜管被冷却套包裹，虽然容易变形，但铜管可以快速更换或矫直。对于圆坯结晶器，一般断面最大尺寸为约230mm^2，或直径430mm，但直径通常小于200mm。大断面管式结晶器和一些小断面管式结晶器壁厚较大，约20mm。（2）板式结晶器。由四块厚度为40~60mm的铜板组装而成。冷面开槽并用钢衬板覆盖，冷却水通过这些凹槽。在另一种设计中，冷却水通过在铜板上机加工形成的环形冷却通道。这些结晶器通常能将窄面调节到不同的宽度，可以在连铸期间操作。

全新的大方坯和小方坯结晶器铜板厚度通常为50~60mm，服役末期厚度大约为40mm。通常在服役期间进行多次维护。

2.2.3.1 结晶器长度

通常的结晶器长度为700mm，长度从500mm到1200mm不等。最近的趋势是900mm结晶器，以便在高拉速连铸时在结晶器出口处增加坯壳凝固厚度。

2.2.3.2 结晶器材质

结晶器材质必须能够迅速将热量从凝固的钢液传递到冷却水，因此需要具备良好的导热性。结晶器多使用铜和铜合金，但必须尽量减少热应力引起的变形。添加银、铬和锆等合金元素是因为它们具有优异的高温性能；图2-5所示的是合金元素添加的详细信息。在某些情况下，结晶器的工作面是（镀）涂层，以尽量减少磨损。据称镀层可以减少铜黏附在凝固坯壳上时形成的星状裂纹，但在许多企业中，尤其是欧洲的企业，在没有镀层的情况下结晶器依然成功运行。

已经开发了各种用镍和铬镀覆结晶器壁的方法。一种技术使用厚镀层，因此结晶器可以在表面维护后重复使用。其他技术使涂层逐渐减薄或使用两段镀覆技术，目的是使结晶器下部的磨损最小。另一种技术镀覆镍铁，结果是硬度增加使耐磨性倍增。结晶器镀层在日本最常见，在其他地方应用有限。

2.2.3.3 结晶器振动

结晶器往复运动的初衷是为了防止坯壳和结晶器之间的粘连，这归功于 Junghans 的工作。除了少数例外，结晶器振动周期是正弦曲线，但在每个振动周期中，有一段时间结晶器向下运动的速度超过拉坯速度。在此期间（称为负滑脱时间或焊合时间），克服了结晶器和坯壳之间的粘连。

结晶器振动可以防止漏钢，在精心控制下，漏钢率几乎为零。结晶器振动源自电机驱动的凸轮，但已开发出液压驱动装置。结构、轴承和杠杆臂的设计是至关重要的，因为行程长度必须在结晶器各处保持相等，并且水平或径向运动要小于 0.2mm。

为获得最佳铸坯质量，振动系统的安装点应与连铸平台和机架分开。有缺陷的振动将导致漏钢率和铸坯表面缺陷增加。最近的工作表明，小焊合时间可以显著改善表面质量。这通常通过小行程长度来实现，在板坯连铸机上行程长度小至 4mm，在方坯连铸机上小至 8mm。与通常 100cpm 或 120cpm 的振动频率相比，小行程结晶器振动频率为 200cpm 或更高。更高频率和更小行程长度的结晶器已经在一些不锈钢连铸机上显现优势，并且变得越来越普遍。为了实现无故障操作，对设计标准和工程标准提出了更严格的要求。

2.2.3.4 调宽结晶器

在过去的十年中，连铸期间板坯连铸机上的结晶器可在线调宽，以响应对不同板坯宽度的需求且连铸不中断。该技术应用于许多板坯连铸机中，通过精心选择的窄板移动顺序，已经实现了 200mm/min 的最大宽度调节速度。

通过在连铸过程中驱动窄面向内或向外小心移动来改变结晶器宽度。在一段时间内进行调整并产生锥形板坯，再加热期间需要特别注意这些锥形坯。图 2-6 显示的是调宽所需的主要部件。在宽度调节期间，窄板锥度的精确控制至关重要，锥度要随宽度的变化而变化。须安装锥度计来连续测量锥度，图 2-7 为调节板坯宽度时窄板移动顺序。

据报道，连铸过程中宽度调整可使产量增加 30%~50%，耐火材料成本降低 30%~50%，收得率增加 0.3%~0.5%，显著节约能源。通过宽度变化增加热装或直接轧制的能力可实现节能，因为这可以匹配轧制进度。

2.2.4 引锭杆系统

开始浇注时必须使用"引锭杆"。引锭杆链由拉辊驱动，引锭头放置在结晶器底部的位置。引锭杆头呈爪状，当钢液进入结晶器时，在"爪"周围凝固，当结晶器充满钢液时，拉坯开始，引锭杆开始从结晶器中拉出部分凝固的钢。当引锭杆头和铸坯的前端离开铸机时，引锭杆头部与铸坯分离，引锭杆链抽出并停在待机位置。对于一个简单的立式连铸机或小截面（小方坯）弧形连铸机，引锭杆可以是一个类似于铸坯横截面的刚性杆。对于更大的截面尺寸和复杂截面连铸机,常使用柔性链式引锭杆(参见图2-8)。不太常见但用于扁平断面铸坯的是板式或带式引锭杆。在这种情况下,薄板贯穿引锭杆的整个长度,并且添加局部包覆件以使引锭杆达到所需的厚度,但其在穿过铸机时仍可弯曲。表2-2所示的是各种引锭杆类型的应用比较。

对于兼用型连铸机,安装多个引导段以匹配多流铸坯,如图2-8所示。引锭杆通常设置成一旦引锭杆脱离铸坯支撑段和拉坯装置就自动脱离铸坯。在实际生产中,引锭杆头部设计成爪形。爪形头部在引锭杆和铸坯穿过铸坯支承辊时结合,然后引锭杆头抬起脱离铸坯。

2.2.5 铸坯支撑系统和二次冷却

从结晶器拉出的铸坯坯壳厚度为 10~25mm(取决于拉速),表面温度从约为 1000℃ 增加至固-液界面的固相线温度(约 1500℃)。坯壳受液态钢液静压力的影响,如果铸坯不经夹持,容易发生鼓肚。

薄的坯壳从结晶器下口出来后需要连续冷却和机械支撑。二冷的喷嘴用于控制冷却,但是水冷保护的铸坯支撑结构也从铸坯中吸收热量。辐射传热也增加了总热量传输。二次冷却系统的设计和操作取决于铸坯支撑系统的类型和设计,而铸坯支撑系统的类型和设计又取决于铸造的断面尺寸和形状。

2.2.5.1 不同铸机支撑系统

铸坯支撑系统在小方坯、大方坯和板坯连铸机之间存在很大差异。对于诸如小方坯角部的约束足以防止坯壳刚离开结晶器时鼓肚。在这种情况下,结晶器足辊与铸坯第一米长度范围左右的每个面上的支撑辊可以提供足够的支撑。这为结晶器下部提供了更广的空间,以便从喷嘴中获得更均匀的冷却。然而,一些小方坯连铸机,在较低的拉速下操作并且生产小于约 130mm 的小方坯或直径小于约 150mm 的圆坯,除了连接到结晶器的足辊之外没有其他支撑。这种铸机的所有辊通常仅用于引导铸坯和复位引锭杆。对于小方坯连铸中的高拉速而言,需要更多的支撑辊。在这种情况下,这些支撑辊间彼此对中及与结晶器出口对中是非常重要的。

结晶器长度通常在 700mm 和 900mm 之间,但是对于一些更高拉速的小方坯连铸机,有时使用结晶器延伸装置。它由四个弹簧负载板组成,通过板中的孔提供喷水冷却。该结晶器与延伸段一起被称为"多级结晶器"。

对于较大的方坯和大方坯,当坯壳凝固壳层薄和高温时,鼓肚倾向加剧,因此支撑辊必须进一步向下延伸。小方坯连铸机和大方坯连铸机的典型支撑系统如图2-9所示。

对于板坯连铸机,板坯宽面的鼓肚延伸到凝固终了,并且宽面的铸坯支撑总是贯穿铸机全长。铸机的后半部分需要拉坯辊。由于板坯连铸机在所涉及的支撑范围和鼓肚力方面是最复杂的,因此铸坯支撑系统的设计和操作的详细说明将集中在板坯连铸机上。应该注意的是,铸坯支撑系统有助于铸坯的冷却,这些冷却影响将在第 2.2.6 节"二次冷却"中提及。

2.2.5.2 结晶器下部支撑系统

目前正在使用板坯连铸机中结晶器正下方的各种铸坯支撑和冷却系统,分别是:支撑辊、冷却格栅、冷却板和步进梁。事实证明，步进梁的结构太复杂，而冷却板的摩擦力太大。

铸坯下部支撑的目的是在最小摩擦力的前提下获得均匀的冷却，同时保持铸坯精确的几何形状。现今支撑辊和冷却格栅常配合使用，使得系统在铸坯和支撑系统之间的摩擦力最小。

结晶器正下方的二次冷却装置很大程度上取决于铸坯支撑系统。例如，由于辊距小，使用辊间扁平喷嘴。然而，对于冷却格栅，使用全锥形喷嘴并对中以将冷却水引导到格栅的矩形孔中。使用冷却板时，引导冷却水通过基体的小孔，在冷却板和铸坯之间产生水膜，提供冷却。

2.2.5.3 铸坯支撑系统

在板坯连铸机中，主要的支撑系统通常包含3~6对支承辊，且能够快速更换整个支撑段。机架通过液压缸联结在一起，并且使用轴承座和垫片预设辊缝。

二冷喷嘴在集水管上对中，使凝固铸坯在轧辊之间的辊缝中冷却。

整个支撑段刚性地固定在连铸机的机架上，轧辊可以通过液压缸调节，以便改变铸坯厚度（通过选择较厚的轴承座）或完全打开。铸机中出现过冷板坯时，必须通过切割移除并定期维护。

有必要使支撑段能够快速更换，通过导轨从铸机中取出每个段（如图2-10所示），吊车沿着导轨从铸机上提起支撑段。在一些铸机中，使用特殊起重机将支撑段水平侧向移除。

已经开发了复杂的有限元模型来预测相邻辊间以及当辊子未与邻辊对中时的凝固坯壳鼓肚程度。这些模型用于设计支撑辊的最佳直径和辊距。轧辊间距必须使得在轧辊接触面之间的鼓肚微不足道。轧辊的直径需要使由液芯产生的钢水静压力和由于轧辊受热不均产生的热应力导致的轧辊弯曲程度最小。

大约在1980年，大多数板坯连铸机使用整体辊，但在过去十年中，分节辊的应用显著增加。整体辊延伸到铸坯的整个宽度上，并由辊的每一端的轴承支撑。随着改进的轴承技术（在高温环境中的冷却和润滑）的出现，大多数新的宽板坯连铸机和许多改造的板坯连铸机现在包含分节辊。分节辊由较短的辊筒组成，每一段通过"中央"轴承横向支撑。这可以更大的范围减小轧辊直径和辊距，同时保持辊刚度并保持辊缝的几何形状。

已经做了很多工作来评估各种辊的设计性能，设计因素和其他操作参数的细节将在后面更全面地讨论。

2.2.6 二次冷却

二次冷却和夹持/拉坯系统从结晶器的底部延伸，直到铸坯完全凝固到切定尺。总的"二次冷却"是几个部分的组合，它们是：辐射冷却、由于喷射的水滴在板坯表面上的蒸发以及积聚在轧辊间的水引起的冷却和传热到轧辊冷却。

在本节中，讨论的细节将集中在喷嘴本身，但这些喷嘴的设计和操作很大程度上由铸坯支撑系统决定，因此不能单独区分喷雾对铸坯凝固的影响。如前所述，在支撑辊之间使用高强度水喷嘴，进一步加速凝固过程并有助于控制和降低铸坯表面温度的波动。

二次喷水冷却系统的作用为：（1）从凝固铸坯中吸收热量；（2）控制喷嘴的设计、布局和水流量，以达到铸坯表面质量所需的最佳表面温度；（3）喷水有助于冷却铸坯支撑辊，尽管支撑辊都有内部冷却。

起初，只有水喷嘴用于二次冷却，但在20世纪70年代末和80年代早期，气雾喷嘴广泛引入。在高压下向喷嘴供水和供气，从而产生更细的水滴粒度，同时还具有大喷射角。这使得喷水均匀成为可能，并且更小的水滴粒度具有增加传热系数的优点。图2-11显示的是所述的两种系统。

撞击在铸坯表面上的水滴应覆盖尽可能宽的区域，但通常由于铸坯支撑系统的存在而变得困难。全锥形喷嘴能够覆盖大的圆形或方形区域，而扁平喷嘴可以覆盖横跨铸坯宽度的区域，但是在相邻辊之间直接喷水冷却时，喷水空间很小。在小方坯连铸机中，主要使用全锥形喷嘴，安装在总集水管上，该总管沿着铸坯的各个面垂直安装。支撑辊位于大方坯连铸机上部以及遍及整个板坯连铸机的长度范围，意味着必须使用扁平喷嘴。在小方坯和小型大方坯连铸机中，整个喷水段的长度在0.5~6.0m，在高拉速板坯连铸机中可以延伸到20m。二次冷却系统沿着铸机的长度分成多个可独立控制的区域。水供应系统完全独立于结晶器冷却水和"封闭"水系统，来冷却轧辊和轴承以及其他元件。

采用气雾冷却时，水滴被高压压缩空气雾化。产生的蒸汽通过大风扇从喷淋室中提取。可能含有水垢和油脂的未蒸发的水沿着铸机下方的水槽返回冷却和清洁设备。

2.2.6.1 喷水冷却

在仅用水进行二次冷却的情况下，水的雾化仅通过水压在喷嘴处产生，而无需来自其他介质的额外辅助。在板坯连铸机中，位于轧辊之间的水平排列喷嘴的数量决定了系统的名称。单喷嘴系统表示一个喷嘴（有时两个）的布置，在每个辊间区域（喷水区）产生广角喷雾（高达120°）；多喷嘴系统涉及在每个喷水区域对许多喷嘴进行分组，喷射角度小。图2-12显示的是这些备用喷嘴系统布置。

单喷嘴系统目前非常适合大多数钢种和尺寸的板坯。它在20世纪60年代中期开始取代多喷嘴系统，因为后者的小喷孔很容易堵塞。与此同时，针对薄板坯和敏感钢种的某些铸机，多喷嘴系统重新应用，喷嘴具有高水流通量配合高拉速。在这种系统中使用的水的悬浮颗粒含量必须很小。

单喷嘴系统的优点是显而易见的：喷嘴更少、供应系统更简单、更易于维护。由于单个喷嘴安装得远离铸坯，因此可以得到更好的保护。广角单喷嘴的另一个重要优点是它们相对较高的流量（相同体积的水，喷嘴较少＝每个喷嘴的流量更大），因此出口孔更大。出口孔决定了喷嘴的流量范围，并且通过控制水压将流量控制在该范围内。然而，大的压力变化也改变了喷射角度，如果压力太低，则喷射角收缩，水喷出喷孔而没有期望的喷射效果。水压力下限通常是0.05~0.1MPa。

单一喷水二冷系统中，喷嘴的共同缺点是它们的体积流量控制范围相对较窄，考虑到连铸设备中常见的工作压力为0.1~0.8MPa（在喷嘴端部），控制范围平均仅为1:3.5。

在连续铸造中，不同钢种的板坯的拉速范围很宽，这种水冷系统喷嘴的有限控制范围可能需要安装两个独立的喷水系统，以便产生合理的水通量。这种系统的特征在于在每个冷却区并排布置两个不同额定流量的喷嘴，并且根据所需的水通量，使用较小的、较大的或同时使用两个喷嘴。当然，这种双系统更加昂贵和复杂。

2.2.6.2 水汽雾化冷却

在水汽雾化冷却系统中，冷却水在混合室中与压缩空气混合，汽水混合物从喷嘴喷出，是高脉冲宽喷射角喷雾。这种汽雾冷却特别适用于易开裂的优质钢。更重要的优点包括特别均匀的冷却模式和非常宽的汽水体积流量范围。

汽-水冷却系统可轻松提供 1∶12 或更高的体积流量控制范围。该系统主要的优点是：(1) 从凝固铸坯中吸收热量；(2) 喷孔的流量大，喷嘴堵塞的可能性很小；(3) 体积流量控制范围大，对于所有钢种和拉速，只需要一种喷嘴；(4) 宽板坯表面（从轧制线到轧制线）冷却水量均匀，铸坯热量散发均匀，降低了铸坯表面局部过冷的危险；(5) 形成极细的水滴，以达到最佳的冷却效果；(6) 细小液滴的有效蒸发使辊缝前积水少。

2.2.6.3 支撑辊

板坯连铸机的支撑辊设计是几个影响因素的调和。在 1980 年以前安装的板坯连铸机，大多数支撑辊是单辊，辊两端有支撑轴承。在 20 世纪 80 年代早期，随着轴承技术的发展来抵抗铸机中的恶劣环境，需要使用 2~3 个分离辊。

所有的支撑辊和轴承都需要水冷，除了铸机上部的一些较小的辊子（使用高二次冷却水流速），所有辊子都是内部冷却的。不同的辊子设计，内部冷却效率可能因设计而异。支撑辊的主要要求是：(1) 辊子直径和间距使铸坯的辊间鼓肚最小化。这还取决于二次冷却程度（即铸坯温度）、拉速（主要决定坯壳厚度）、铸坯下方的距离和钢种。钢的蠕变性能根据钢种变化而有很大差异。在 12m 的弧形连铸机上，切点处的钢水静压力为 $86t/m^2$，因此作用在凝固坯壳上的力非常大。鼓肚程度也与时间有关，凝固坯壳从一个辊子传递到下一个辊子所需的时间与拉速有关；(2) 辊子的几何外形应保持稳定。如果轧辊直径太小而且长 2m（典型的板坯单辊长度），轧辊会因以下原因而弯曲：1) 铁水静压力；2) 在运行过程中不对称的温度分布导致的轧辊热应力；3) 在铸坯停止期间，不对称的温度分布显著加剧。

水冷支撑辊本身可从凝固铸坯中吸收大量热量，并且吸收的热量取决于轧辊设计。各种类型的轧辊设计和轧辊冷却方法在图 2-13 中给出，图示是辊的设计和冷却方法。图例是整体辊，但许多原理也同样适用于分离辊。

由于外围孔和螺旋冷却水通道靠近表面，因此辊表面较冷，被称为"冷"辊设计，而中心钻孔冷却被称为"热"辊设计。"冷"辊比"热"辊能吸收更多来自铸坯的热量。然而，冷辊更稳定，铸坯停止时不易发生永久弯曲，并且不易于发生由于严重不对称的温度分布导致轧辊弯曲产生严重的热应力。如果辊子中心处的永久弯曲大于约 1mm，可能导致铸坯内部质量变差。

在几何稳定性和吸热能力方面，相关人员做了大量工作评估辊子性能。值得注意的是，喷水量会影响辊子吸收的热量。在水不进入辊缝的情况下，吸热量分别为 44kW/m 辊和 26.5kW/m 辊。还获得了关于各种类型辊的几何稳定性的数据。凸度计已用于测量辊子状态和铸坯的鼓肚。这些测量计由位于铸坯末端的线性位移传感器（LDT）组成，这些传感器刚性地固定在铸机中，其中 LDT 适当地安装在辊子的背面或铸坯表面上。

在铸坯中的任一位置均使用三个这样的凸度计,两个在相邻的辊上,一个在两个辊之间的铸坯上。这些仪器可以长时间留在铸坯中,并且研究了针对许多情形(例如铸坯停止或减速)以及铸造不同钢种的各种二次冷却条件下辊子和铸坯的状态。需要避免辊子几何外形的偏差,因为这会导致铸坯表面和内部质量变差。稍后将更详细地描述。

上述工作都是在整体辊上进行的,这些辊必须在足够小的直径和辊间距的前提下进行调整,以防止铸坯的辊间鼓肚。辊需要足够大的直径,避免在机械和热负荷下弯曲,以保持良好的辊缝几何形状。如前所述,在过去的十年中分离辊的应用显著地增加。大多数新的宽板坯连铸机和许多改造的板坯连铸机现在都包含分离辊。这意味着单个辊长度大大减小,这显著减少了辊的弯曲,因此可以实现更小的辊直径和辊间距。

辊缝几何形状也会受到轧辊磨损的影响。因此,辊材料也是非常重要的,辊材料和有效冷却的组合可以减少辊磨损,辊磨损是辊缝变形的原因。辊材料还需要耐烧裂和应力腐蚀开裂,为了满足这些要求,辊子涂覆了包含12wt%Cr和88wt%Fe的"硬面"金属层。

2.2.7 铸坯矫直和拉坯装置

弧形连铸机铸坯弯曲运行,在水平出坯前矫直。矫直机的设计(或在立式结晶器中浇注后,铸坯弯曲的顶弯段)取决于连铸机半径,铸坯截面尺寸,待浇注的钢种和其他连铸参数。细节将在下面的第2.2.7.1节中描述。另外,需要赋予铸坯足够的动力和牵引力,使得拉坯是可靠和连续的。

2.2.7.1 铸坯矫直

如前所述,弯曲的铸坯需要被矫直以实现水平出坯。矫直装置的设计取决于若干因素,重要的是确保矫直产生的应变引起的应力小于铸坯的极限强度。

完全或部分凝固的铸坯上的应变可以由标准梁弯曲理论确定,但是由于涉及温度的蠕变,因此设计矫直铸坯所需的总应变,应变速率也是一个重要的因素。铸坯的应变分布还取决于铸坯是完全凝固还是仍然存在液芯,这两种情况将分开处理。在需要更高产量的现代连铸机中,矫直过程中通常存在液芯。

A 铸坯完全凝固

在这种情况下,应变分布完全取决于初始曲率和铸坯厚度,如图2-14所示。表面应变:

$$\varepsilon_s = \frac{b}{2R} \times 100\% \qquad (2-8)$$

这是铸坯上表面的拉伸应变和下表面的压缩应变。通过在多个弯曲点上施加所需的应变,甚至在给定长度的铸坯上连续矫直,可以降低应变速率。稍后将描述这些系统。

B 带液芯矫直

在这种情况下,上部和下部凝固壳都看作是单独的梁,理论应变取决于凝固边缘的夹辊。当凝固坯壳达到理想的厚度并且坯壳的形状受到二维传热的影响,夹辊的影响在低的长宽比下可能是显著的。因此采用两种方法,分别称为"软盒"和"硬盒"法。

软盒法:当上部和下部凝固壳彼此独立地变形时,即当没有凝固边缘的约束时,该铸坯被看作"软盒"。这是板坯的情况,其中长宽比高并且坯壳厚度与板坯宽度相比较小。

图 2-15 显示了由于在切点处矫直而在凝固壳中产生的应变分布。

由于温度梯度的影响,这种中性轴沿着上坯壳和下坯壳中心线的假设并不是很严谨。有限元分析表明,真正的中性轴更靠近冷面。在铸坯上部外表面和下部坯壳的固/液界面处都存在拉伸应变。这些应变是矫直时的铸坯半径和坯壳厚度的函数。

在这种情况下,表面应变由下式给出:

$$\varepsilon_s = \varepsilon_i = \frac{t}{2R} \times 100\% \tag{2-9}$$

式中,ε_s 为外表面应变(A 处的拉伸;D 处的压缩);ε_i 为固/液界面应变(C 处的拉伸;B 处的压缩);t 为坯壳厚,m;R 为铸机半径,m。

硬盒法:在这种情况下,铸坯弯曲主要受凝固边缘的刚度影响,假定中性轴沿断面中间厚度的延伸,并且表面应变类似于铸坯完全凝固的情况,即:

$$\varepsilon_s = \frac{b}{2R} \times 100\% \tag{2-10}$$

在这种情况下,固/液界面的应变由下式给出:

$$\varepsilon_i = \frac{b-2t}{2R} \times 100\% \tag{2-11}$$

已经证实,"软盒"方法适用于具有高长宽比的板坯或大方坯。"硬盒"方法仅适用于方坯和小断面大方坯。

如前所述,应变速率通常决定是否会发生裂纹(内部或表面)。特别是在固/液界面处的钢的强度在矫直温度下非常低,但是在这些温度下蠕变会导致由应变引起的应力迅速降低。因此,通过降低应变速率,可以将应力保持在低值,并且可以扩展矫直段来实现总的高应变。这是通过使用多点矫直来完成的。

在极限情况下,在冶金长度为 L 的铸机上使用连续矫直。在这种情况下,应变率(min^{-1})如下:

$$\dot{\varepsilon}_s = \frac{\varepsilon_s v}{L} \tag{2-12}$$

式中,v 为拉速,m/min;L 为连续矫直装置的长度,m。

2.2.7.2 铸坯弯曲

在使用立式直结晶器的情况下,铸坯在结晶器下方弯曲到适当的半径。此时凝固坯壳仍然相对较薄,因此应变通常不如用带液芯矫直时那么高。原理相同,并且许多带有直结晶器的铸机使用多点弯曲来实现所需的铸坯半径,同时降低应变率以避免内部缺陷。在这种情况下,需要使弯曲辊的未对准最小化以减小应变。

2.2.7.3 拉坯装置

在恒定和可控的条件下从铸机中拉出铸坯,需要施加足够的动力和牵引力。拉坯力必须足以克服作用在铸坯上的摩擦力。摩擦力可能来自:结晶器中的铸坯摩擦、载荷导致支撑辊与轴承的摩擦、轧辊之间的铸坯鼓肚产生的滚动摩擦。还应注意的是,铸坯本身的自重有利于减小所需的拉坯力。图 2-16 所示的是大方坯连铸机和板坯连铸机的拉坯装置。

板坯连铸机的拉坯装置是多辊拉坯，牵引力和动力分布在多个辊对上。使用稍微超过该位置处的钢水静压力的液压力，对驱动辊对施加合适的牵引力。在板坯连铸机中产生的拉坯阻力只能通过多辊拉坯系统克服。这种系统在早期成功地降低了铸坯的拉坯阻力，随着铸坯沿连铸机前进，将阻力降低到低水平。

结晶器下方的拉力由于铸坯的自重而轻微减小。拉力保持在相对较低的值，直到铸坯到达矫直部分突然增大。铸坯完全凝固后，由于消除了钢液静压力，拉力增加的速度减小了。

2.2.8 铸坯输送和后续处理

铸坯完全凝固之后，切定尺切割，将其输送到下一工序。该过程再次从连续操作变为间歇操作。该间歇操作必须设计成使得出料的周期比将铸坯从切割装置输送到后续系统中的短。否则，先前操作的连续性将被中断。

铸坯剪切装置尽可能靠近铸坯支撑的末端或最终凝固点。除了可以通过剪切机切割的小方坯或薄板坯外，其他铸坯切割都通过割枪完成。使用气体切割是因为切割铸坯是最经济的方法。然而，气体切割有一个小的"切口"损失。该切口损失大约为 8~12mm。对于小方坯或薄板坯，可以使用剪切机；这些装置更贵，但金属损失为零。

通过专门设计的喷嘴中氧气和燃料气体燃烧（通常为丙烷或焦炉煤气）来实现切割。氧气纯度通常至少为 99.5%。一个或多个喷嘴接近铸坯的边缘，加热边缘一小段时间以预热铸坯，然后开始切割操作。喷嘴以预定的速率移动，穿过铸坯直到它被切断。对于兼容型连铸，通常单独操作板坯切割的火焰切割机以切割铸坯。当应用兼用型铸造时，应该在板坯切割处安装 3 个切割机。

测量系统用于穿过切割机的铸坯，也用于铸机位置，以便计算切割铸坯的正确位置。在许多情况下，在铸机上安装双燃烧喷嘴，为了控制质量，可以切割硫印或蚀刻的铸坯横截面。将样品送到相邻的实验室进行分析。对于小方坯和薄板坯连铸机，可以使用剪切机。这些设备通常太昂贵而不适用于较大的截面尺寸。剪切机的主要好处是它们不会因切割而造成切口损失。

3 连铸传热和凝固

3.1 结晶器传热

水冷铜结晶器的传热细节、传热机理和凝固行为是连铸工艺中最重要因素。结晶器通过一定程度的控制,以尽可能均匀的方式从钢中吸收热量。铸坯的表面质量主要依赖于结晶器的相关参数,因为这是铸坯表面在结晶器中形成,因此可能是许多表面缺陷的发源地。均匀的传热也有助于避免漏钢。

图 3-1 给出了凝固坯壳和冷却水之间的温度分布。热通量 Q 由下式计算:

$$Q = h_{ss}(T_{ss} - T_{hf}) = \frac{K}{D}(T_{hf} - T_{cf}) = \frac{h_{cf}(T_{cf} - T_{bw})Kw}{m^2} \tag{3-1}$$

式中,h_{ss} 为凝固坯壳传热系数,$kW/(m^2 \cdot K)$;T_{ss} 为凝固坯壳外表面温度,℃;T_{hf} 为铜板"热面"温度,℃;T_{cf} 为铜板"冷面"温度,℃;K 为铜板导热系数,$kW/(m \cdot K)$;h_{cf} 为"冷"面传热系数,$kW/(m^2 \cdot K)$;T_{bw} 为冷却水温度,℃;D 为铜板的厚度,m。

从结晶器中的钢液温度来看,凝固坯壳层上存在温降,这将在后面讨论。坯壳和结晶器壁的热面之间的润滑膜和铸坯收缩形成的间隙,是影响铸坯和结晶器冷却水之间热通量的主要因素。结晶器铜板高导热性确保了铜板上的温度小幅下降。由于水冷通道中存在热边界层,结晶器壁的冷面显著高于整体冷却水温度。然而,该边界层可受冷却通道中冷却水流动条件的影响,并且根据充分验证的传热理论,可以相当笃定地预测跨越边界层的温降。必须保持冷却水的流速足够高(8m/s),以避免核态沸腾。

铸坯凝固壳层和铜板之间的界面为主要组成热阻部分,是一个复杂的现象,需要更详细地讨论。这很大程度上受所用润滑剂类型的影响。

在小方坯边长小于约 130mm 和圆坯直径小于约 130mm 时,难以使用耐火浸入式水口。在这些情况下,使用定径水口"敞开式"浇注,并采用惰性气体保护,从弯月面上方的铜板中的小孔供给菜籽油作润滑剂。图 3-2(a)显示了使用菜籽油作润滑剂时结晶器的细节。

在板坯和大方坯连铸中,浸入式水口(SEN)与保护渣一起使用,保护渣在结晶器中的液态渣和钢液之间形成流体熔渣。图 3-2(b)显示了使用浸入式水口和合成保护渣时结晶器中的细节以及与铜板的界面。

使用结晶器保护渣替代菜籽油的主要优点是:

(1)浸入式水口(SEN)与结晶器保护渣一起使用,是一种更有效的保护浇注方法。

(2)可以防止结晶器中钢液表面的辐射热损失,并防止因钢液表面凝固导致的"镀层"缺陷。

(3)保护渣吸收上浮至结晶器熔池中的非金属夹杂物(例如 Al_2O_3)。

(4)保护渣提供到铜壁更均匀的传热。

结晶器保护渣的成分和物化性需要保证来自钢液的热量产生足够厚度的熔渣层，并且黏度使得熔渣能够连续地流入铜板壁处的弯月面。

另一个基本要求是结晶器以正弦方式振动，使得在振动周期中的一段时间内，结晶器将以比铸坯更快的速度向下运动。

图 3-3 为振动周期内结晶器比铸坯向下运动更快的周期部分，被视为润滑（甚至摩擦）和均匀传热之间的折中。

结晶器振动、结晶器保护渣供料和结晶器液面变化之间的相互作用非常复杂，已经开发了几种计算机模型来确定结晶器保护渣消耗率和弯月面处的凝固特性，作为结晶器振动、结晶器液面和结晶器保护渣特性的函数。所有这些因素决定了渣膜厚度，而渣膜厚度又决定了界面的热阻。另外，凝固坯壳和铜板之间的间隙受到表面温度和坯壳收缩的影响，可能导致形成气隙，气隙的特征取决于铸坯断面尺寸和形状。图 3-4 显示的是各种断面铸坯在坯壳和结晶器壁之间形成气隙，如在弯月面下方形成的气隙。

这些气隙也可以沿着结晶器的长度方向变化，通常从弯月面以下开始增加，可以通过小方坯和大方坯横截面的锥度来抵消。在板坯结晶器中，只有窄面跟随横截面收缩，因此只有窄板是锥形的。由于铸坯鼓肚，板坯的宽面没有形成间隙，并且宽面彼此平行。

通过嵌入结晶器铜板的热电偶测量结晶器铜板的热通量做了大量研究。这项工作一般聚焦热通量、热通量分布、结晶器壁温度、所用结晶器润滑剂的类型、钢种成分和生产实际之间的相互关系。

结晶器壁温度和沿结晶器长度的热通量分布的影响因素包括：(1) 流速和冷却水的速度；(2) 使用的结晶器润滑剂类型；(3) 铸造钢种的碳含量；(4) 拉坯速度。

(1) 冷却水流量的影响。冷却水流量范围的变化幅度很宽。对于冷却水流量的这种宽泛的变化，热通量是相对恒定的，这证实了从凝固坯壳中吸收热量的最大限制因素是钢与铜板的热面之间的边界层。可以看出，边界层冷却水流速低导致铜板温度升高。

(2) 结晶器润滑的影响。图 3-5 显示了连铸保护渣和菜籽油结晶器润滑剂对结晶器热通量分布和热面温度的影响。可以看出，采用菜籽油时，热通量相当高，尤其是弯月面区域的热通量非常大。

(3) 碳含量的影响。钢种成分，特别是含碳量对整个结晶器传热的影响已见诸报道。图 3-6 是钢中碳含量为 0.02%~1.6% 结晶器中的平均热通量。碳含量对传热的影响导致一些质量问题在碳含量 0.06%~0.14%（包晶范围）内更加突出。

对于含碳 0.1% 的碳钢，已经观察到沿结晶器长度的厚度不均匀的坯壳。有人提出，坯壳厚度和传热的不均匀性是由 γ-δ 相变和在该碳含量水平上发生的体积变化和收缩引起的。图 3-7 是靠前的热电偶测量的两个碳含量水平的温度，从这些记录中可以看出，温度波动在碳含量较高（0.55%）的钢中较低，而含碳 0.15% 的钢在连铸期间出现明显的温度波动。

(4) 拉坯速度的影响。拉坯速度对结晶器中的热通量分布和平均热通量也有显著影响。图 3-8 显示的是在 0.8~1.3m/min 的不同拉坯速度下，平均热通量与结晶器距离的关系。

结晶器铜板的温度分布：图 3-9 是结晶器壁中的垂直温度场以及测量数据。这些数据通过许多大型企业测量获得，且在铸坯凝固时用作结晶器中的边界条件。

3.2 二次冷却区传热

通常,二次冷却系统包括一系列区域,每个区域负责铸坯凝固行进时的控制冷却。喷淋的介质是水或空气和水的混合物。

3.2.1 二冷区凝固

坯壳厚度沿所谓二冷区的末端增加,支撑辊变得更大并且间隔更远。二冷区长度通常也称为冶金长度,因为这是铸坯凝固和铸态组织形成的阶段。根据铸坯的横截面和拉坯速度,它可以长 10~40m。各个区域喷嘴的水流通常是计算机控制的,并且随着连铸条件的变化自动调节。

在铸坯通过最后一对支撑辊之后,进入出坯辊道,由一个或两个氧乙炔火焰切割机切割。

根据菲克定律可以有效地预测坯壳的生长:

$$L = v\left(\frac{D}{K}\right)^2 \tag{3-2}$$

式中,D 为坯壳厚度;L 为到结晶器弯月面的距离,凝固始于弯月面;v 为拉速;K 为经验常数,主要取决于钢种和连铸机机型。

上式也可用于计算铸坯完全凝固的铸造长度(L)(即没有剩余液芯)。

3.2.2 二冷区传热

二冷区的传热包括喷雾冷却、自然对流、热辐射、支撑辊接触等,是非常复杂的过程。喷雾冷却是传热中最重要的过程。

二冷区有三种基本的传热形式:

(1) 辐射。辐射是二冷区上部的主要传热形式,由下式描述:

$$Q = \sigma \varepsilon A (T_s^4 - T_a^4) \tag{3-3}$$

式中,σ 为修正的 Stefan-Boltzmann 常数;ε 为辐射系数,通常为 0.8(衡量自身相对于完美的散热器或黑体的辐射能力);A 为表面积;T_s,T_a 分别为铸坯表面和环境温度。

(2) 热传导。当铸坯通过轧辊时,由于相对接触,热量通过坯壳传导并且通过轧辊传递。傅里叶定律描述了这种传热形式:

$$Q = \frac{kA(T_i - T_o)}{\Delta x} \tag{3-4}$$

式中,k 为坯壳的导热率;A,Δx 分别为坯壳的横截面积和厚度,热量通过横截面和厚度方向传递;T_i,T_o 分别为坯壳的内外表面温度。如图 3-10 所示,这种形式的传热也发生在夹持辊中。

(3) 对流。对流传热可以描述为:来自喷嘴的水滴或气雾高速运动,穿过钢表面的蒸汽层然后蒸发,带走热量。由牛顿冷却定律描述:

$$q = hA(T_s - T_w) \tag{3-5}$$

式中,h 为传热系数,常数,通过确定水通量、喷嘴类型、喷水压力(和空气压力),最后通过实验确定钢表面温度;A 为表面积;T_s,T_w 分别为钢表面和喷水温度。

具体而言，喷淋室（二次冷却）传热具有以下功能：增强并控制凝固速度，促使铸坯在该区域完全凝固；通过调节喷水强度调节铸坯温度；冷却铸机夹持段铸坯。

夹持段是二冷区的组成部分。一系列夹持辊夹持铸坯，延展至相对的铸坯面，也可能夹持铸坯边缘。该区域重点提供铸坯导向和夹持，直到凝固坯壳可以自主支撑。

为了避免铸坯质量缺陷，必须使得辊和未弯曲铸坯的应力最小化。因此，仔细选择辊布置，包括间距和轧辊直径，以使辊间鼓肚和液/固界面应变最小化。

铸坯支撑需要保持铸坯形状，因为铸坯本身是凝固坯壳包着的液芯形式，该液芯具有与铸坯高度相关引起的铸坯鼓肚钢液静压力。最值得关注的区域是铸机中的高位。这里，鼓肚力相对较小，但是坯壳较薄并且强度最低。为了弥补这种固有的弱点并避免坯壳破裂和由此产生的漏钢，辊径和辊间距尽可能小。在结晶器正下方，通常通过足辊支撑铸坯的四个面，而在铸机下部的区域仅支撑铸坯的两个宽面。

3.3 铸坯的凝固

3.3.1 凝固过程

铸坯凝固从水冷结晶器中开始。在结晶器中形成包含液芯的坯壳，当铸坯通过由大量支撑辊对组成的铸坯引导段时，液芯逐渐凝固。使用水雾和辐射冷却的组合在二冷区中完成始于结晶器弯月面处的凝固过程。连铸技术的凝固特性源于连铸工艺的动态特性。特别涉及：（1）处理结晶器中非常高的热通量；（2）形成初始坯壳，尽量避免在结晶器运动过程中开裂；（3）根据钢种的凝固动特性设计连铸参数，以消除铸坯中的表面和内部缺陷。

在凝固期间，固/液界面可能是稳定的也可能是不稳定的。对于纯金属，界面移动速度通过界面扩散的潜热来控制。稳定界面的特征源于来自界面的潜热通过固体的快速传导。界面上的任何扰动都会遭遇液相并且导致界面消失。因此，沿着界面的所有位置将以相同的速度移动。另一方面，对于不稳定的界面，液相温度低于熔点。液相过冷，导致热量从界面通过液体散出。因此，界面上的任何扰动都会出现在过冷的液相上，从而使更多的热量从界面散出，并且界面上的扰动将使界面树枝状晶或侧枝比其他位置生长更快。

在合金凝固过程中不稳定的界面更常见，因为对于合金而言，界面不稳定会导致界面附近产生溶质浓度差。在纯金属中，高的冷却速率可以在凝固发生之前，将熔体的温度降低到远低于熔化温度，即产生过冷。过冷度可达 T_m（熔体的完全熔化温度）的 10%～30%，T_m 取决于冷却速率和熔体的纯度。

3.3.2 结晶器中钢液的凝固

结晶器的主要功能是使得铸坯进入二冷区前生成具有足够强度的坯壳。主要影响因素是形状、坯壳厚度、均匀的坯壳温度分布、无缺陷的内部和表面质量、孔隙率最小、非金属夹杂物少。

为了理解结晶器中的初始凝固现象，有必要更广泛地了解凝固。金属的凝固是一个复杂的过程，在宏观上涉及固体金属、液态金属和结晶器的传热，随着凝固的进行，固相和液相之间具有连续移动的界面。在微观上，凝固涉及形核、传热、界面能和偏析。描述凝固现象所需的数学分析是复杂的,并且通常需要通过迭代计算。然而,可以回顾相关理论和方程,加深对凝固的基本理解。

结晶器由无底的水冷铜板或圆管组成。铜板刚性地固定在钢支撑结构上,并且具有一系列相邻的冷却水缝,用于冷却结晶器。结晶器振动为了防止粘钢。为了进一步促进润滑,结晶器保护渣或植物油注入铸坯和结晶器之间的间隙,固相保护渣控制传热。整个装置设置在一系列用于弯曲和矫直铸坯的辊道上方,并且喷水冷却铸坯。

结晶器中的凝固是决定铸坯质量的关键。对于板坯连铸机,通过多孔浸入式水口将钢液浇注到结晶器中。添加的结晶器保护渣熔化在钢液面形成液态渣。结晶器保护渣是氧化物和其他添加剂的混合物,含有氧化铝、二氧化硅、石灰、碳酸钠和某些氟化物。结晶器保护渣用于保护钢不被氧化、润滑铸坯并控制传热。结晶器保护渣与结晶器表面接触而冷却,钢液接触冷却的保护渣,快速凝固并且形成铸坯的初始坯壳。随着连铸的继续,坯壳向下运行并凝固。结晶器的长度通常约为 1m,并且钢的停留时间从小方坯连铸中的 1min 到在板坯连铸的约 20s。停留时间不足以使铸坯完全凝固,但是足以生成坚固的坯壳以便在铸机的二冷区中夹持。如果坯壳不够厚,无论出于何种原因,都会发生漏钢。在板坯离开结晶器后,直接在表面上喷水,并通过夹辊传热而进一步冷却。结晶器内部的凝固铸坯如图 3-11 所示。

在连铸过程中,钢液与结晶器表面接触传热,温度降低到液相线以下,使初始钢液凝固。因此,该方法通常首先形成与结晶器接触并包围剩余液体的凝固壳。

如果凝固中断,凝固前沿产生周期性的厚度不均匀,具有几厘米量级的波动。然而,如果凝固继续进行,则由于不断增厚的壳,凝固前沿形态较少依赖于结晶器-坯壳界面,不均匀性趋于消失。这种微结构对随后的凝固过程是有害的,并且与铸坯开裂有关。Cisse 等人已经发布了铸造过程中热机械生长不稳定性的实验结果。有人提出,结晶器表面适当的几何形状可能有助于促进铸坯坯壳均匀生长,从而防止坯壳厚度不均匀。

Murakami 等人提出结晶器表面的周期性沟槽,由于钢液的不充分润湿而导致沿着结晶器-坯壳界面的间隙尺寸可控,从而做出许多重要的改进。最显著的改进是结晶器-坯壳间更均匀的接触,并且由于更慢但更均匀的吸热而减少了铸坯的裂纹。为了验证该假设,使用具有机加工沟槽的结晶器对铝合金进行重复试验。研究的一个关键参数是沟槽间距或波长及其对坯壳厚度不均匀性的影响。有人提出,有可能进行振痕选择工艺,系统可"摘除"结晶器表面振痕或振痕带,使坯壳生长得更均匀。

对于光滑的结晶器表面(或至少没有明显的周期性凹凸),由于各种与工艺相关的条件和材料特性/冶金转变,散热的波动是由结晶器-坯壳界面热通量的随机变化引起的。凝固早期,坯壳厚度不均匀表明:随着坯壳变厚,结晶器-坯壳边界条件对凝固前沿的不规则生长的影响减小。在具有纯正弦曲线分布的结晶器表面的理想情况下,限制因素是表面几何形状,因为这在散热截面中产生空间扰动。

连铸的早期凝固以与结晶器钢液接触的弯月面区部分凝固的形式发生。在凝固铸坯下降期间防止初始薄壳的黏结和撕裂是结晶器的主要功能之一。最大限度地减少坯壳黏结和撕裂取决于坯壳强度,且铸坯表面和结晶器壁之间的摩擦力必须在临界值以下。通过引入润滑的结晶器振动,实现了坯壳的摩擦和拉坯阻力的最小化。

在近终形连铸工艺早期凝固阶段,从熔体到结晶器的总热阻主要取决于金属/结晶器界面的热阻，与传统的连铸相反，凝固坯壳和润滑膜的热阻也很重要。在双辊连铸中，钢液直接在没有润滑剂的基板（两个水冷辊）上凝固。

铸坯的最大拉坯速度受结晶器内凝固壳的生长和强度的限制。

连铸坯壳厚度的近似见式（3-5），凝固坯壳的厚度 S 与凝固时间 t 的平方根成比（t 也有重要影响）：

$$S = K\sqrt{t} \tag{3-6}$$

K 的"平均"值为 26，其中 S 以 mm 为单位，t 以 min 为单位。在实际计算中，23~32 范围内的值用于简单的补偿，适用于不同钢种、铸坯尺寸、喷水速率的影响等。

因此，总凝固时间由 K 确定，对于矩形铸坯横截面，由铸坯厚度确定。厚坯需要比薄坯更长的凝固时间。对于给定的铸机长度，拉坯速度随厚度的平方减小。这对铸机设计有直接影响，至少在板坯连铸中，必须防止铸坯鼓肚。换句话说，铸坯厚度和拉坯速度决定了"冶金长度"。

在连铸工艺中，凝固过程显著影响钢材质量。钢液从中间包注入结晶器后，沿结晶器壁凝固并形成坯壳。坯壳厚度不断增长，变成坚固的板坯或薄带。在数学建模中，成形过程也是至关重要的，因为液相与固相共存并且凝固持续进行。钢液流动和温度分布决定了钢液凝固和微观组织。因此，凝固过程对钢的质量和产率产生显著影响。成形过程中适当的流量控制对于优质钢生产至关重要。

3.4 连铸模拟

经调查，钢铁厂运营成本高昂，在设计、故障排除和优化工艺时使用所有手段是明智之举。例如，物理建模、水模拟等可以深入了解钢液的流动规律。连铸过程的复杂性和控制环节使得模拟难以进行。然而，随着计算机硬件和软件功能的不断增强，数学建模正成为全面理解连铸的重要工具。

3.4.1 物理模拟

先前对连铸中钢液流动的理解主要借助物理水模型。水模拟试验中新配置问世前，此技术是测试和理解钢液流动的有效方法。

物理模型的构建基于通过匹配控制连铸的重要现象的几何和力的平衡来满足模型和实际过程之间的某些相似标准。为了用水模型再现钢液流动模式，两个系统主导力的比率必须相同。这确保了模型和实际工艺的速度比在每个位置都相同。定义了铸流中的一些重要的力比，它们定义了一组无量纲组。无量纲组的大小表示两种力的相对重要性。可以忽略比值非常小或非常大的无量纲组，但连铸过程中所有其他的无量纲组必须在物理模型中实现匹配。

必须选择合适的几何尺度和流体来实现这些匹配。幸运的是，水和钢具有非常相似的运动黏度（μ/ρ）。因此，雷诺数和弗劳德数可以通过构建全尺寸水模型同时匹配。满足这两个标准，足以在等温单相流动系统的建模中得出准确结果，例如连铸水口和结晶器的模拟已经取得了巨大的成功。实际上，任何几何尺度的水模型对于大多数流动系统都能得到合理的结果，只要两个系统中的流体流速足够高，产生完全湍流和非常高的雷诺数。由于通过中间包和结晶器水口的流动是重力驱动的，因此在这些系统的任何水模型都满足弗劳德数，其中液压头和几何形状都按相同的比例放缩。

物理模型必须满足热相似。例如，在钢包和中间包稳定流动的物理流动模型中，相对于主要的惯性驱动流动，热浮力较大。因此必须保持弗劳德数在模型与实际工艺中相同。在钢包中，难以估算流速，但可以将雷诺数的平方除以修正的弗劳德数，称为格拉索夫数。惯性力在结晶器中占主导地位，因此可以忽略热浮力。热浮力的相对大小可以与全尺寸热水模型中的一致，例如，通过控制温度和热损失，使得 $\beta \Delta T$ 在模型和铸机中都是相同的。然而做到这点并不容易，因为控制热损失取决于诸如流体导热率和比热以及容器壁导热率等特性，这些特性在模型和实际中是不同的。在其他系统中，例如涉及低速、瞬态或凝固的系统，同时满足许多对传热而言十分重要的相似标准实际上是不可能的。

当物理流动模型用于研究其他现象时，还必须满足其他力比。例如，对于包含粒子运动的研究，重要的是匹配涉及惯性、阻力和浮力的力比。通过匹配终端浮选速度等其他条件来满足，即：

$$v_{\mathrm{T}} \equiv \frac{g(\rho - \rho_{\mathrm{p}}) d_{\mathrm{p}}^{2}}{18\mu(1 + 0.15Re^{0.687})} \tag{3-7}$$

式中，v_{T} 为粒子终端速度，m/s；ρ，ρ_{p} 为钢液、颗粒密度，kg/m³；d_{p} 为粒径，m；μ 为钢液黏度，kg/(m·s)²；g 为重力加速度，9.81m/s²；Re 为粒子雷诺数，$Re = \rho v_{\mathrm{T}} d_{\mathrm{p}}/\mu$。

例如，在全尺寸水模型中，可以使用密度为 998kg/m³ 的 2.5mm 塑料珠来模拟钢中 2300kg/m³ 的 100μm 实心球形夹杂物，因为它们具有相同的末端上浮速度（式（3-7）），更容易想象。

有时，不可能同时匹配所有重要标准。例如，在研究两相流动时，将气体注入钢液，发生的新现象变得重要。流体密度取决于当地的气体分数，因此满足相似需要额外匹配气体分数及其分布。必须增加水模型中使用的气体分数，以便考虑将气体注入钢液时发生的大约 5 倍的气体膨胀。还必须对局部压力进行调整，这也会影响这种膨胀。除了匹配气体分数外，气泡尺寸应该相同，因此也应该匹配涉及表面张力的力比，例如韦伯数。在试图实现这一点时，可能需要偏离注入点的几何相似性并对模型表面上蜡以修改接触角，以便控制初始气泡尺寸。如果气体动量不可忽视，例如高气体注入速率，那么气体和钢液密度的比率也必须相同。为此，水中的氦气是钢中氩气的合理匹配。在许多情况下，同时匹配所有重要的力的比值非常困难。在某种程度上，通过近似水模型也可以揭示真实过程。

3.4.2 计算模型

近年来，降低计算成本和增加商业建模软件功能使得计算模型用作理解复杂过程（例如连续铸钢）的工具更加有效。计算模型具有易于扩展到其他现象的优点，例如传热、粒子运动和两相流，这对于等温水模型而言是困难的。它们还能够更准确地表示钢液的流动情况。例如，不需要物理干扰水模型底部的出口，并且可以考虑移动的凝固坯壳。

数值模型现在可以模拟大多数对连铸重要的现象，包括：

（1）保护渣到铜板壁更均匀的传热；

（2）受氩气泡、热力和溶质浮力影响的复杂几何形状（水口和熔池）中的完全湍流瞬态流体运动；

（3）保护渣、钢液内部及两者界面间的热力学反应；

（4）钢液和上浮至钢液表面上的固体保护渣层内的流动和传热；

（5）钢液表面和界面的动力学运动，包括表面张力、振动和重力感应的影响，以及几个阶段的流动；

（6）通过湍流钢液散热，降低过热；

（7）溶质传输；

（8）通过钢液传输的复杂几何形状的夹杂物，包括浮力、湍流相互作用的影响，以及夹杂物可能位于水口壁、气泡、凝固坯壳壁和钢液面上；

（9）凝固弯月面、固体渣边缘、熔融保护渣渗透、钢液、保护渣层和夹杂物颗粒之间的弯月面区域的热、流体和机械相互作用；

（10）通过凝固坯壳、坯壳与结晶器之间的界面（包括保护渣层和生长的气隙）和结晶器传热；

（11）保护渣沿坯壳与结晶器之间的间隙流动；

（12）结晶器壁和支撑辊的变形与磨损；

（13）晶体的形核，包括在熔体内和在结晶器壁上的形核；

（14）坯壳的凝固，包括枝晶、晶粒和微观组织的生长，相变、析出和微观偏析；

（15）热收缩、相变和内应力导致的凝固坯壳收缩；

（16）外力（结晶器摩擦、支撑辊之间的鼓肚、拉坯力、重力）、热应变、蠕变和塑性（随温度、钢种和冷却速率而变化）在凝固坯壳内产生应力；

（17）裂纹；

（18）在微观尺度和宏观尺度上的耦合偏析。

这个过程惊人的复杂性导致不可能同时将所有现象建模在一起。因此，有必要做出合理的假设，并去除或忽略不太重要的现象。定量建模需要结合影响研究目标的所有现象，因此每个模型都需要具有特定的研究目标。一旦选择了控制方程，它们通常被离散化并使用有限差分或有限元方法求解。重要的是进行适当的数值验证。在求解非线性方程时，数值误差通常源于过于粗糙的计算域或不完全收敛。解决已知的测试问题并进行网格细化研究以得到与网格无关的解决方案是帮助验证模型的重要方法。最后，必须检查模型与实验室和企业实验的测量结果，然后才能对参数研究的实际过程进行定量预测。

3.4.3 流体流动模型

流体流动的数学模型可以应用于连铸过程的许多方面，包括钢包、中间包、水口和结晶器。典型的模型解决了以下连续性方程和不可压缩牛顿流体的 Navier Stokes 方程，它们基于计算域中每个点的质量（一个方程）和动量（三个方程）守恒：

$$\frac{\partial v_i}{\partial x_i} = 0 \tag{3-8}$$

$$\frac{\partial}{\partial t}\rho v_j + \frac{\partial}{\partial x_i}\rho v_i v_j = -\frac{\partial P}{\partial x_j} + \frac{\partial}{\partial x_i}\mu_{\text{eff}}\left(\frac{\partial v_i}{\partial x_j} + \frac{\partial v_j}{\partial x_i}\right) + \alpha(T_0 - T)\rho g_j + F_j \tag{3-9}$$

式中，$\partial/\partial t$ 为对时间的微分，s^{-1}；ρ 为密度，kg/m^3；v_i 为 x_i 方向的速度分量，m/s；x_i 为坐标方向，x，y 或 z，m；P 为压力，N/m^2；μ_{eff} 为有效黏度，$kg/(m \cdot s)$；T 为温度场，

K；T_0 为初始温度，K；α 为热膨胀系数，m/(m·K)；g_j 为 j 方向的重力，m/s^2；F_j 为其他力，例如电磁力；i,j 为坐标方向。

式（3-9）中倒数第二项解释了热对流对流动的影响。最后一项解释了其他力，例如施加的电磁场力。这些方程的解决方案形成了在每个点处产生压力和速度分量，通常是三维向量。在这些过程中涉及的高流体流速下，这些模型必须包含湍流流体。要做到这一点，最简单但计算要求最高的方法是使用足够精细的网格来捕捉所有湍流漩涡及其随时间的运动。该方法被称为"直接数值模拟"，用于在板坯连铸机的结晶器中产生瞬时速度场。在150万节点的网格上30s流量的模拟需要在SGI Origin 2000超级计算机上进行30天的计算。将计算结果与水模型中流动粒子的速度测量结果进行比较。这些计算结果显示，钢液流动模式的结构对瞬态事件很重要，例如间歇捕获夹杂物颗粒。

结晶器中的流动具有重要意义，因为它影响许多对铸坯质量具有深远影响的重要现象。这些现象中的一些在图 3-12 中示出。它们包括通过撞击凝固坯壳的钢液射流（和弯月面处的温度）消耗过热、液面保护渣层的流动和夹渣、液面轮廓和液面波动，以及皮下夹杂和气泡。需要调和这些设计，以同时满足避免每一个缺陷生成互相矛盾的要求。

重要的是将模拟扩展到工艺上游，以便为目标区域提供足够的入口边界条件。例如，结晶器中的流量计算前应先计算通过浸入式水口的流量。水口几何形状极大地影响结晶器中的钢液流动并且易改变，因此它是建模的重要主题。

随着氩气注入速率的增加，钢液流动模式发生根本变化，这需要解决附加的气相方程，以及需要气泡尺寸的知识。通过施加电磁力也可以改变钢液流动模式和钢液混合，电磁力可以制动或搅拌钢液。这可以通过麦克斯韦方程、欧姆方程和电荷守恒方程来建模。这些现象增加到耦合模型方程的巨大复杂性使得这些计算结果不易得出，但这是正在研究的主题。

4 连续铸钢工艺

4.1 引言

钢液成分和温度满足要求并去除非金属夹杂物后，钢液（热量）运输到钢包回转台并倒入中间包，在连铸机上凝固以获得铸坯。

引锭杆插入结晶器底部开始连铸工艺。钢液一旦进入结晶器，就会凝固在无底铜结晶器的水冷壁上，形成急冷坯壳。结晶器垂直振动，防止坯壳粘连到结晶器壁上。连铸机驱动辊连续地从结晶器中拉出坯壳，拉坯速度与进入结晶器的钢液流相匹配，因此连铸在稳定状态下理想地运行。

连铸中最关键的部分是弯月面的凝固，凝固发生在坯壳顶部与结晶器和钢液表面相交处。这是铸坯表面形成的地方，如果出现液面波动的问题，则可能形成诸如表面裂纹和夹杂物之类的缺陷。

在结晶器出口下方，薄的凝固壳（6~20mm 厚）充当容器以支撑剩余的钢液，构成铸坯内部。喷水或气雾冷却支撑辊之间的铸坯表面。在铸坯中心完全凝固后，将铸坯用氧乙炔火焰切割机切割成所需长度。

在铸坯离开铸机一定距离后，将其加热至均匀的温度并轧制成板材、棒材、轨道钢和其他的标准形状的坯料。

4.2 钢液处理

出于生产效率和质量的考虑，现代炼钢有一种趋势，即将温度调节、脱氧和合金化等耗时的操作转移到钢包处理站。在涉及连续铸造工艺的情况下，这些处理尤其重要，因为必须严格控制钢液温度和成分。

4.2.1 温度控制

连铸工艺中钢液进入结晶器时的温度控制比传统铸造工艺要求更精确。过高的过热度可能导致漏钢并产生树枝晶，这通常与铸坯内部质量差相关。另一方面，温度太低可能堵塞水口进而导致铸造困难，产品洁净度差。

板坯中间包钢液的温度通常高于液相线温度 5~20℃，小方坯或大方坯连铸高于液相线温度 5~50℃；温度的差别取决于钢种，例如，对于小型不锈钢板坯连铸，差值约为 45℃。

为了在整个连铸工艺中将钢液温度保持在规定的范围内，钢包中的钢液温度均匀性是至关重要的。在连铸之前需要搅拌以破坏钢包中的温度分层。搅拌已成为一种常见均匀温度的做法，即通过钢包底部的多孔塞或在单独的清洗站通过中空塞棒注入氮气或氩气。

图 4-1 说明了如何将大多数炉次 X70 钢的温度限定在钢包目标温度的 +7~0℃ 范围内。

4.2.2 成分控制

可以在真空或渣洗处理期间进行钢液成分的控制。在对样品或电动势进行分析的基础上，均匀钢液后测量氧活度，可以修正铁合金添加量以确保的合理脱氧。在搅拌熔池时快速（惰性载气保护渣喷吹、喂丝或喷丸）添加微调脱氧剂。通过钢包扒渣，减少对合金的需求，简化调整过程。

真空处理是通用的手段，若钢包冶金效果良好，不需要强制使用。不过，低真空处理是在连铸之前深脱氢或深脱碳的唯一手段。

脱氧后的钢液也可以使用钙基和镁基化合物脱硫。这些元素在适宜条件下添加时不仅具有良好的脱硫能力（在镇静钢中深加入），而且它们还将剩余的硫转化为非塑性夹杂物。此外，添加钙基和镁基化合物有利于去除氧化物。钙处理的另一个优点是将氧化铝夹杂物转化为液态铝酸钙，减少水口堵塞。

为了保持钢包冶金的优点，需要特别注意后续钢液二次氧化。必须特别注意的是减少卷渣、包衬材料、钢种及减少铸流卷入空气。

4.2.3 夹杂物控制

夹杂物会在铸坯中产生许多缺陷。例如，低碳铝镇静钢由于塑性差而易发翼缘裂纹，轴承钢有疲劳寿命问题。塑性和疲劳寿命都易受到钢中硫化物和氧化物夹杂的影响。星状裂纹沿着钢带表面平行于轧制方向呈线状分布。星状裂纹对汽车用低碳铝镇静钢薄板有严重危害，导致表面缺陷和成形性差等问题。夹杂物通常由来自脱氧的铝酸盐产物或由来自夹带的结晶器保护渣中的复杂的非金属夹杂物组成。

在钢包浇注期间或在添加脱氧剂之前测量出钢氧含量。对于 IF 钢生产，出钢氧含量通常很高，从 250ppm 到 1200ppm 不等。

铝脱氧产生大量 Al_2O_3 夹杂物。这表明应限制洁净钢的出钢氧含量。但是，如图 4-2 所示，出钢氧含量和钢的洁净度之间没有相关性。这与在大量添加铝后形成氧化铝簇状夹杂物，其中约 85% 的夹杂物容易上浮到钢包渣中，剩余夹杂物尺寸小于 30μm 的说法一致。

当然，能否忽略出钢氧含量取决于夹杂物的上浮时间以及钢包精炼的效果，两种手段均可以去除大部分的夹杂物。图 4-3 显示了 RH 处理如何达到相同的最终出钢氧含量水平，无论初始出钢氧含量如何，只要脱气时间足够长，比如 15min。最后的考虑因素是出钢含氧量会影响脱碳率，特别是生产超低碳钢。

钙处理将低碳铝镇静钢中的氧化物和硫化物变为液相，并改变它们在凝固坯壳中的形状和塑性。液态铝酸盐比固态氧化铝夹杂簇更容易聚集和上浮。除了避免水口堵塞之外，还有利于将它们移除到渣中并降低全氧含量。为了获得液态夹杂物，钙含量要适当。钙含量的合理范围很窄并且取决于氧化铝含量，如平衡相图所示。此外，硫含量必须低，使钢液中的液态夹杂物超过铝镇静钢中铝含量，如图 4-4 所示。由于钙具有很强的反应性，因此只有在钢脱氧后或少渣量的前提下加入效果才好。

4.3 中间包到结晶器

将钢液从钢包通过中间包浇入无底的水冷铜质结晶器中。结晶器的底部由作为其活底的引锭杆封堵，开浇后引导铸坯从结晶器连续运行到拉矫机。结晶器中开始凝固的铸坯在到达拉矫机之前通过二冷区系统冷却。引锭杆到达分离装置与铸坯脱钩分离。之后，可以用火焰切割或机械剪以与热铸坯相同的速率移动定尺切割。

中间包通过使用塞棒、滑动水口或其他控流装置将确定流量的钢液浇入一个或多个结晶器中。根据中间包耐火衬的性质，中间包初始温度不尽相同。结晶器不仅形成铸坯断面形状，而且还导出定量的热量，使得铸坯坯壳足够坚固，以便在到达结晶器出口时进行正常运行。结晶器由紫铜或铜合金制成，这取决于铸坯的形状和断面尺寸。通常，管式结晶器用于较小断面铸坯。结晶器的内表面可以镀有铬或钼，以减少磨损并适应来自铸坯的传热。结晶器是倒锥形的，以适应不同钢种的收缩。目前使用的结晶器总长度在400~1200mm，但是其长度通常是在700~800mm。通常通过相对于铸坯正弦振动结晶器并添加润滑剂（油或连铸保护渣）以减弱结晶器和钢之间的摩擦来改善坯壳黏结的问题。润滑剂，特别是连铸保护渣，具有额外的冶金功能。润滑剂的选择取决于铸坯质量和浇注条件，保护渣选择铸坯质量匹配特别重要。

结晶器中的钢液面可以手动控制或通过自动系统控制。任何一种方法都可用于保持液面恒定或与浇入结晶器的钢液速率相匹配，即适应拉速的变化。通过中间包塞棒或通过改变浇注速率实现手动控制。自动控制系统可以通过结晶器壁上的探针测量放射性、红外辐射或温度测量，以确定钢液面并通过启动塞棒机构（用于稳定浇注速率）或控制拉坯辊（改变拉速）来应对液面变化。

连铸的引锭杆的类型取决于安装类型。刚性引锭杆可用于垂直连铸系统，而铰接式引锭杆或柔性带式引锭杆必须弓形安装。引锭杆可以以不同的方式连接铸坯，一种是通过使用可溶于引锭杆的连接元件（平板、螺钉或轨道钢碎片）焊接钢液；另一种方法是将钢液浇注在引锭杆特殊形状的头部中，使其能够通过解锁与铸坯分离。

离开结晶器时凝固坯壳的厚度首先取决于钢与结晶器接触时间长短，但也取决于结晶器的热导率和钢液的过热量。可以使用式（4-1）相当准确地确定坯壳厚度：

$$C = K\sqrt{t} \tag{4-1}$$

式中，C 为坯壳的厚度，mm；K 为凝固系数，$mm/min^{1/2}$；t 为凝固时间，min。

凝固系数取决于操作条件，结晶器内和结晶器附近坯壳的凝固系数在20~26之间。坯壳厚度取决于拉坯速度，离开结晶器时凝固坯壳的厚度约为铸坯厚度的8%~10%。结晶器下的二冷加速了凝固。冷却剂通常是水，但有时也使用水/空气混合物或压缩空气。二冷区分为几个区域以适应冷却水流速。通过喷嘴将所需的水喷射在整个铸坯上。相对于铸坯横截面和拉坯速度，钢液静压力非常高，以至于必须支撑铸坯以防止弯曲。相关设备在生产大方坯的企业中是昂贵的，板坯生产设备尤为昂贵。

在敞开浇注中，钢液直接穿过空气从钢包流到中间包或从中间包流到结晶器。在这些情况下，未受保护的铸流从空气中吸入氧气（和氮气），并在钢液中形成有害的夹杂物。这些夹杂物被转移到结晶器中，在结晶器中它们被保留在铸坯内或上浮在钢液的表面。存在于钢液表面上的那些夹杂物随后或被捕获在凝固壳中，导致产品在轧制中的产生表面缺陷或在结晶器下方的壳中发生灾难性漏钢事故。除了在暴露的钢流中直接形成夹杂物之外，夹带在铸流中的空气也可以在结晶器和中间包中与钢液反应。

为了避免这些问题，采用注流保护浇注。由于问题的严重性，首先强调保护中间包和结晶器之间注流。然而，现在广泛使用钢包至中间包注流保护，特别是在铝镇静钢板坯铸造中，防止氧化铝夹杂物是至关重要的。保护浇注由两种基本方式，同时有许多变化和组合：(1) 气体保护；(2) 耐火管保护。

由于耐火管保护操作困难，气体保护经常用在小方坯连铸机，因为小方坯连铸机上没有足够的空间来使用耐火管而且不能结晶器壁和耐火管之间发生金属凝固。

耐火管有各种各样的设计，包括：Pollard 钢管护罩，其中气体以低速从管的中点引入，并在管和水口之间，以及管和结晶器之间流出；使用弹性联轴器完全封闭中间包和结晶器之间的注流；截棱锥形外壳；液氮隔断（如图 4-5 所示）。使用氮气或氩气作为保护气体。单独的气体保护通常不用于防止钢包到中间包注流氧化。有一种圆环设计，一端连接在钢包上，当钢包朝向中间包下降时，另一端砂封，从而形成一个封闭的箱子；然后向箱子里加氩气。

耐火管通常用于铸造铝镇静钢，用在钢包和中间包之间，以及中间包和结晶器之间的注流保护。当结晶器对应的中间包盛满钢液时，管的一端连接到钢包（或中间包），另一端浸入钢中。耐火管通常由熔融石英或氧化铝石墨制成。

耐火管的设计很重要，特别浸入钢中的出口端。一种是直通式设计。通常在结晶器中使用另一种类型，即多孔（开口）设计，例如双侧孔耐火管，其中管的底部封闭，并且两个侧孔位于管的底部附近。这种类型的耐火管避免了注流深入到凝固界面中并且改变结晶器中的钢液流动模式。因此，注流中的夹杂物不会留在凝固部分中，而是上升到液态金属的表面，并被由结晶器保护渣形成的熔渣除去。

在许多企业中，耐火管附属设备的设计需要考虑更换能力，从而可以在不中断的情况下连铸。在一些企业，将氩气注入耐火管中以避免因运动金属流的文丘里效应导致孔和接头吸入空气。

4.4 二次冷却

4.4.1 二冷控制

在现代板坯连铸机中，二次冷却、铸坯夹持和拉坯形成紧密集成的互锁系统，该系统还包括铸坯弯曲和矫直。在较老的小方坯和大方坯连铸机机型中，这些组件具有更广泛的功能和并且可以分离。

二冷系统用于生产具有适当形状、内部质量和表面质量的产品。为了实现这些目的，离开结晶器的铸坯在喷水区域中冷却，并通过轧辊夹持和拉坯，直到凝固铸坯到达切割机和水平送料台。

4.4.2 铸坯夹持

铸坯支撑包括由带有液芯的铸坯形状的坯壳的约束。由钢液高度产生的静压力使铸坯鼓肚，特别是铸坯刚好在凝固壳厚度较薄的结晶器下方（图 4-6）。铸坯的四个面通常都在结晶器下方被支撑，在铸机下部，只有两个面被支撑。除了钢液静压力和坯壳厚度之外，轧辊间距还受铸坯表面温度和钢种影响。

铸坯弯曲和矫直除了夹持铸坯之外，从垂直面到水平面引导铸坯通过预设圆弧的一系列夹辊必须足够坚固以承受弯曲反作用力。在顶弯期间，凝固坯壳的外弧处于拉伸状态，内弧处于压缩状态。由此产生的应变是弧形半径和铸造的特定钢种强度的函数，这是至关重要的；外弧过大的应变会导致表面缺陷（裂纹）。为了最小化表面缺陷同时保持最小有效圆弧半径，引入了三点弯曲（即三个弧具有逐渐变小的半径）。

在弯曲完成之后安装多辊矫直机，顾名思义，多辊矫直机即将铸坯矫直并将铸坯方向从垂直变到水平。在矫直期间，铸坯被矫直成"不弯曲的"水平方向，期间铸坯上表面受力由压缩力变为张力，铸坯下表面受力由张力变为压缩力。

4.4.3 拉坯

通过驱动辊将铸坯拉过连铸机的各个部分。驱动辊系统设计用于在铸坯表面产生压力，以提高表面质量。因此，目标是通过连铸机"推动"铸坯，而不是用会产生表面缺陷的拉伸应力"堆垛"铸坯。在任何情况下，驱动辊施加的压力不得过大，压力过大会使铸坯断面变形。

矫直之后，将铸坯从辊台上输送到切割机，在切割机中铸坯被定尺切割。然后将铸坯分组或直接输送到精轧机，或者在小方坯的情况下，主要输送到步进梁式冷床，以保持产品的平直度。

4.4.4 冷却水系统

通过强制对流的冷却水来控制结晶器中的传热，冷却水系统必须适应凝固产生的高传热速率。通常，冷却水进入结晶器底部，穿过位于外结晶器壁和钢套之间的一系列平行水通道，从结晶器顶部排出。

冷却水系统的主要控制参数是：
(1) 在所需水温、压力和铸坯质量下的水量。
(2) 通过结晶器衬套周边通道的冷却水流速均匀。

4.4.4.1 水量、温度、压力和质量

通常采用加压再循环的闭环系统。水流速率应足以吸收来自铸坯的热量，而不会过多地增加总体水温。温度的大幅升高可能导致传热效率降低和结晶器温度升高。出于同样的原因，结晶器的进水温度也不应过高，还需要适当的结晶器水压。如前所述，较高的水压倾向于抑制沸腾，但过高的水压可能导致结晶器变形。

水质是结晶器内衬上水垢沉积的重要因素。水垢沉积可能是一个严重的问题，因为它在结晶器冷却水界面处产生额外的热阻，这增加了结晶器壁温度，导致诸如蒸汽和衬板强度降低的不利影响。水垢的类型和数量主要取决于冷却水的温度和速度、结晶器的冷面温度和水处理的方式。

4.4.4.2 水流速度

为了获得适当的流速，冷却系统水流速度足够高，以有效地传热。流速太低会产生更高的热阻，这可能导致沸腾，带来不利影响。通常，冷却水速度越高，结晶器温度越低。冷却系统还应在结晶器周围流速分布均匀，并使直接水冷的面积最大化。通过集水管和窄

板水通道的适当设计可以实现均匀的流量分布。

监控结晶器冷却系统的参数可以对铸造过程评估。例如，在恒定的冷却水流速下，从结晶器表面散发的热量直接与入口和出口水温 AT 之差相关。因此，过大的 DT 可能表示一个或多个结晶器面的异常低的流速，而过小的 DT 可能表示一个或多个结晶器面的异常过热。相对面的不同 DT 可能是由于不对称的结晶器变形或铸坯未对准造成的。

4.4.5 拉速控制

连铸工艺、拉坯速度和连铸机性能决定了设备的产量。由于每个连铸机的限制条件，如钢种、连铸机长度和连铸机复杂程度不尽相同，拉坯速度差异很大。

图 4-7 是板坯厚度和拉坯速度之间的散点关系，这些散点关系揭示了技术的快速发展。虽然对较低拉坯速度的钢材难以分析，但对要求严格的产品具备优异的性能。很明显，板坯连铸机未来的生产率必须高于一般过去认为的可能性。

在图 4-8 和图 4-9 中，图中绘制的线做如下计算。

对于小方坯/大方坯连铸机（图 4-8），偏差比板坯连铸机小得多，拉坯速度的范围是板坯连铸机的一半。如前所述，值得注意的例外与美国钢铁公司有关。

轨道钢坯的拉坯速度受到限制，以防止断裂和表面裂纹。虽然板坯连铸机的统计数据表明拉坯速度有进一步发展的巨大潜力，而拉速很高的大方坯和小方坯连铸机可能因漏钢过于频繁而无法使用，如果拉坯速度显著增加将不得不开发新技术。

考虑到作为铸坯横截面函数的吨/分钟/铸坯的生产率（图 4-9），由于相似厚度的较大横截面积，因此板坯连铸机的吨位自然要大得多。这些数据是根据提供的统计数据得出的最小、平均和最大生产速率计算得出的。

虽然大型转炉可以满足板坯连铸机（以最大速度运行）的生产率，但是技术开发非常困难；然而，6~8 流连铸机仍以较小的转炉进行钢液供应。

4.5 连铸的开浇、控制及问题

开浇包括将引锭杆（基本上是金属梁）封闭结晶器的底部，将钢液浇入结晶器中，钢液一旦凝固，就用引锭杆拉坯。

许多操作现在完全由计算机控制。钢包长水口、中间包和结晶器上的电磁和热传感器检测钢液面、钢液的流速和温度，并控制拉坯辊的速度来改变铸坯的拉坯速率。水口的钢液流量由顶部的滑动水口控制。计算机还可以设定结晶器振动速率和结晶器保护渣添加速率，以及结晶器冷却速率（通过控制水流量）。

虽然大量的自动化控制有助于生产没有收缩和很少偏析的铸坯，但如果钢液不洁净或在铸造过程中变"脏"，自动控制的作用就微乎其微。钢液变脏的主要原因之一是氧化，这在钢液的熔融温度（高达 1700℃）下迅速发生。为防止氧化，钢液应尽可能地与大气隔离。这就是为什么用水口传输钢液，以及为什么用保护渣覆盖结晶器熔池顶部的原因。中间包钢液通常也用渣层覆盖。

连铸中可能出现的主要事故是漏钢。当铸坯的薄壳破裂时，铸坯内仍然熔化的钢液溢出并弄脏连铸机，停机成本十分昂贵。通常，漏钢是由于拉坯速度太高或者钢液温度太高，坯壳没有凝固到所需的厚度。这意味着凝固在矫直辊终了，并且铸坯受矫直过程中施加的应力可能发生断裂。

4.6 结晶器保护渣

结晶器保护渣定义为在连铸结晶器中添加到钢液顶部的保护渣或颗粒,部分熔化形成接触钢液的熔化层,使钢免受再氧化、吸收非金属夹杂物、在坯壳通过结晶器时润滑坯壳并控制从凝固坯壳到结晶器的传热。保护渣粒度、形状和堆密度将影响保护渣的流动性。

当结晶器保护渣添加在结晶器中钢液的顶部时,它必须流过并完全覆盖暴露的钢液表面,这在使用保护渣自动给料装置时尤其重要。如果结晶器保护渣不易流动,则钢液将暴露在空气中。结果是绝热缘不足和钢的再氧化增加,结晶器保护渣吸收钢液中非金属夹杂物的能力降低。

4.6.1 结晶器保护渣的特性

4.6.1.1 结晶器保护渣的选择

结晶器保护渣主要是合成渣,连铸过程中添加在钢液面上。被钢液熔化的保护渣通过结晶器的振动流入铸坯和结晶器之间的间隙中。为了满足铸坯质量要求,必须选择合适的保护渣,因为保护渣类型会影响振痕深度和结晶器保护渣消耗(影响铸坯润滑)。对于结晶器保护渣消耗的几种估算公式可以在文献中找到,其中大多考虑了保护渣的黏度和结晶器振动设置。一个近似公式是:

$$Q_s = \frac{A}{\eta^{0.5}} \cdot \frac{1}{v_c} \cdot t_N + B \tag{4-2}$$

式中,Q_s 为结晶器保护渣消耗量;η 为保护渣的黏度;A 和 B 为拟合参数,具体取决于实际的配置。

4.6.1.2 结晶器保护渣的种类

结晶器保护渣可分为多种类型。最常见的类型是:

(1)粉煤灰保护渣:机械混合物,主要成分是粉煤灰。粉煤灰缺乏均匀性,限制了这种保护渣的使用。

(2)合成保护渣:细粉料机械混合物。

(3)预熔或烧结保护渣:这类保护渣显著的特点为预熔和粒度比例。

(4)颗粒状保护渣:球形或挤出颗粒的低粉尘保护渣。球形颗粒特别适用于自动保护渣添加。

结晶器保护渣为特定钢种和钢厂专门设计。实际的结晶器保护渣化学成分根据所需的性能变化很大,典型成分范围如表4-1所示。

4.6.1.3 保护渣的主要功能

(1)隔热。保护渣必须提供隔热以防止搭桥和浮渣。保护渣的绝热性提高了钢的弯月面区域的温度,这有助于使振痕减缓并且可以减少诸如气孔的亚表面缺陷。隔热性能的主要限制因素是未反应的保护渣的密度,但所用的碳的类型和保护渣的物理性能也会影响隔热性能。保护渣体积密度太低可能产生扬尘。这个问题通过粒状保护渣来解决。由于其颗粒形状，球形颗粒保护渣比其他类型的保护渣（例如保护渣和挤压颗粒）具有更好的流动性。

（2）化学隔离。如果保护渣中可被还原的氧化物含量低，则熔渣层可有效地将钢液与大气隔离，从而防止钢的氧化。

（3）吸收夹杂物。液态渣可以吸收从钢液中上浮的非金属夹杂物，例如氧化铝夹杂。图 4-10 显示的是保护渣的碱度（CaO/SiO_2）如何影响保护渣吸收氧化铝的能力。将氧化铝棒浸入具有不同碱度的液态渣中。对于不同的保护渣，将棒的直径（ΔR）差对时间作图。正常保护渣碱度为 0.8~1.25。非金属夹杂物的吸收随着碱度的增加和渣中 Al_2O_3 的减少而提高。保护渣较低的黏度改善了夹杂物捕获和溶解的动力学。应该注意到，这种保护渣的性能是在生产洁净钢和保障浸入式水口寿命之间的折中。

（4）润滑。润滑可能是结晶器保护渣最重要的功能。保护渣必须起到在凝固壳层和水冷结晶器内壁之间的润滑膜作用。较低的保护渣黏度或凝固点能提供更好的润滑，且有助于防止黏结。

碱度（CaO/SiO_2）增加时，或 F 或 Na_2O 含量增加，将增加保护渣重结晶的趋势，如图 4-11 所示。

通过对结晶器中的保护渣样品中测量凝固保护渣横截面的不透明区域来建立图 4-12 中所示的结晶指数。指数 0 表示保护渣完全是玻璃状，而指数 3 表示保护渣是 100%不透明的。不透明区域的 X 射线衍射证明它是晶体。由于保护渣趋向于结晶，黏结显著增多。

（5）匀热通量。最后的要求是提供均匀的热通量。这对于防止坯壳的不均匀凝固是必要的，坯壳的不均匀凝固会导致铸坯的开裂。中碳钢在凝固后具有较大的收缩率，这使得这些钢种特别容易开裂，如图 4-13 所示。

具有较高凝固温度的保护渣在坯壳和结晶器之间的保护渣膜中产生较厚的结晶层，降低了传热速率。结晶器热流的减少如图 4-14 所示。

"新"保护渣具有更高的凝固点和更低的黏度。表 4-2 显示的是使用"新"高凝固温度保护渣改善纵向表面裂纹。

4.6.1.4 保护渣物化性质

黏度、凝固点、熔点和结渣速度通常被认为是保护渣最重要的性质。对于某些钢种，有时需要特别考虑保护渣密度或化学性质。

黏度是保护渣消耗的主要因素，较低的黏度导致保护渣消耗增加。

Al_4O_3 的增加对保护渣黏度和凝固温度的影响如图 4-15 和图 4-16 所示。

保护渣 C 和 D 的黏度略微升高，因此在使用过程中更稳定。添加 MgO 可有效地稳定黏度，因为 Al_2O_3 被吸收。

洁净钢生产中，熔融保护渣中 Al_2O_3 的增加量通常低于 3%，源自良好的保护渣脱氧和覆盖技术。黏度和凝固温度的综合影响润滑和传热。结晶膜比玻璃膜更多孔，这是减少传热的一个因素。结晶器涂层的类型、气隙和钢液氢含量也影响热流的传输速率。

保护渣主要由基础渣料、助熔剂和含碳材料（熔速调节剂）组成。添加氧化物对保护渣的黏度和凝固点的影响如图 4-17 和图 4-18 所示。

保护渣特性的变化是一般性的,并且与保护渣化学成分有关。保护渣熔点不仅受保护渣化学成分的影响,还受矿物组成影响。

碳对保护渣熔化速度、烧结倾向、绝热性能和渣圈有显著影响。保护渣熔化特性受碳类型的影响(由于燃烧温度的差异)。结晶器保护渣中的细碳颗粒量有助于确定其熔化模式。

细碳颗粒较多的 α 型熔化模式在半熔渣层中形成离散的液滴。这些保护渣迅速熔化,特别适用于高速和不稳定的结晶器条件下的连铸。粗碳颗粒较多的 β 型熔融模式形成部分烧结的半熔渣层。这些保护渣缓慢而稳定地供应,并且已经展现出适合于在低或中间拉坯速度下铸造裂纹敏感钢的优势。

碳含量不合理的保护渣(保护渣、烧结渣或颗粒)在拉速变化期间可能不具有足够的保护渣深度。图 4-19 显示了具有正确数量和类型的碳的保护渣,在拉速变化期间和拉速变化之后可保持足够的钢液保护渣深度。

4.6.2 连铸工艺条件对结晶器保护渣的影响

拉坯速度、振动周期和结晶器设计是影响结晶器保护渣设计的一些关键参数。

(1)拉坯速度。随着拉坯速度增加,保护渣消耗量减少,如图 4-20 所示。如果保护渣消耗太少,可能会因润滑膜变薄而导致坯壳与结晶器壁黏结。

低拉坯速度也可能产生负面影响。降低钢的流速会导致结晶器中出现低温点,这可能导致浮渣或晶间搭桥。浮渣可能会导致诸如裂纹或气孔等质量问题,而完全搭桥会导致粘连。需要特别注意保护渣的隔热性能,以防止在低拉坯速度下出现问题。

(2)振动周期。高振动频率和短行程减小了振痕深度,同时有助于消除横向开裂问题。较高的振动频率减少了负滑脱时间,与高拉速减少保护渣消耗效果类似。因此,振动周期的变化同时需要调节保护渣以防止黏结。

(3)结晶器设计。通过采用结晶器涂层减少星形裂纹,来改善板坯表面质量。由于涂层提供的耐磨表面,结晶器寿命也延长了。由于涂层剥落问题,镀铬经常被镍基材料取代。然而,镍对结晶器的热流有显著的负面影响,需要添加改善传热的保护渣。

4.6.3 钢种对结晶器保护渣的影响

(1)低碳铝镇静钢(LCAK)(C<0.08%)。LCAK 钢具有良好的高温力学性能,不易发生裂纹。然而,提高产率需要提高拉坯速度。这些钢种的要求是良好的表面质量和内部质量,并防止黏结。可通过具有良好隔热性能、优异的非金属夹杂物吸收性、良好的润滑性和稳定性的保护渣实现。保护渣性能稳定,吸收 Al_2O_3 能力强,同时对黏度没有不利影响是非常重要的,特别是在较高的拉坯速度下。保护渣较低的黏度或凝固点有助于保护渣在较高的拉坯速度下提供足够的润滑。

(2)中碳钢(MEDC)(C 0.08%~0.18%)。由于与包晶凝固相关的铸坯收缩增加,这些钢种易开裂。防止纵向和横向裂纹是至关重要的。这是通过减少通过结晶器的热流并具有特定熔化速度的保护渣来实现的。

凝固点高的保护渣可有效减少表面裂纹,有助于减少通过结晶器的热通量,而低黏度使保护渣能够提供足够的润滑。

(3) 高碳钢（HC）（C>0.18%）。这些钢种的特征是高温强度差，铸造温度低，并且通常在较低拉速下浇注。对保护渣的要求是减少表面浮渣、防止气孔和黏结。这是通过具有良好隔热性能、正确选择碳类型和良好润滑的保护渣实现的。低黏度和熔点使保护渣获得快熔化速度。低保护渣密度和适当的碳添加有助于实现良好的隔热，以防止过早凝固。

(4) 超低碳钢（ULC）（C<0.005%）。与 LCAK 相比，这些钢种在凝固过程中具有较窄的糊状区。这增加了由凝固前沿的快速移动引起的皮下缺陷的风险。向钢中添加钛会产生化学反应，从而改变结晶器保护渣的性质。保护渣必须设计成可提供良好的表面质量、良好的内部质量和防止黏结。通过提升的非金属夹杂物的吸收能力、最小碳的吸收能力、改进的隔热性能、良好的润滑形、性能稳定和最小的夹渣或附着的保护渣来实现钢种的要求。

高碱度以及特殊的氧化物增加了 Al_2O_3 的吸收能力并减少了夹渣和黏结。当与稳定的结晶器条件相结合时，通过保持足够的保护渣深度可以最小化碳吸收，降低保护渣的固定碳含量也是可行的。表 4-3 列出了每种钢种的保护渣的物理性质。

5 铸坯缺陷和质量控制

5.1 前言

连铸工艺并非不存在缺陷。许多研究者做了相当大的努力来建立连铸机的合理设计、操作和维护,以确保最终产品的冶金质量。

避免常见的铸造缺陷,可以在连铸工艺中生产的优质铸坯。在连铸过程中可观察到一些缺陷。例如,当结晶器出口温度低于钢液的熔点时,发生铸坯表面粗糙和裂纹等表面缺陷。随后,在结晶器内发生凝固并生成粗糙的铸坯表面,可能导致铸坯破裂。

当钢液温度过高时,凝固前沿远离结晶器出口,可能导致漏钢,中断连铸过程。此外,当凝固前沿进入冷却装置时,冷却装置入口可能刮擦铸坯表面并形成含有蒸汽泡的位置,蒸汽泡因冷却水接触高温铸坯表面而产生。随后,在铸坯表面出现周期性凹槽。这些凹槽可能因蒸汽泡破裂时,汽泡速度高和中心的高压对铸坯表面造成冲击而产生。该过程符合在铸坯表面周期性凹槽产生期间听到的周期性敲击声。然而,这些凹槽的周期性可能是由于膜内的周期性干燥/润湿转换或在冷却装置的入口处的过渡沸腾区所导致。

波纹型表面缺陷在凝固前沿开始形成,在铸造过程中系统受到机械振动时产生。

熔池内的压头影响工艺的连续性和铸坯尺寸,压头必须等于钢液从熔池行进到没有压力的结晶器出口处的损失。该压头必须与拉坯速度和铸坯尺寸适当协调,以补充铸造过程中消耗的金属。否则,过大的压头会导致结晶器出口处的钢液漏钢,而压头不足将导致铸坯的横截面减小或产生变化。

铸坯质量受连铸工艺的影响。决定铸坯表面质量的关键因素是凝固前沿的位置,凝固前沿必须位于结晶器出口处或其附近。凝固前沿位置受工艺的影响,例如结晶器出口温度、冷却速率、结晶器-冷却器距离和拉坯速度。例如,增加拉坯速度导致将凝固前沿位置移至结晶器出口,反之亦然。拉坯速度的增加必须通过冷却速率的增加或结晶器-冷却器距离的减小来实现。执行操作的准确性或者说干预过程的连续性,需要大量的个人技能和经验。

连铸缺陷包括夹杂物、裂纹和形状缺陷。

(1) 夹杂物因钢种和脱氧方法以及由连铸机操作员使用的设备及受机型影响的拉坯速率而异。夹杂物的控制在很大程度上取决于操作。

(2) 外部和内部裂纹通常根据其几何外观进行分类。除了某些钢种外,内部裂纹可能会在轧制过程中重新焊合,减少质量问题。它们被称为半径中线和中心线裂纹。

(3) 表面裂纹分为纵向、中面和角部裂纹、横向中面和角部裂纹以及星形裂纹。

(4) 形状缺陷。根据形状和尺寸的不同,铸坯可分为板坯、小方坯、大方坯和圆坯等。铸坯的形状缺陷不仅是物理形态问题,还与裂纹、漏钢等其他问题有关。

5.2 铸坯洁净度

5.2.1 铸坯夹杂物及其来源

5.2.1.1 夹杂物的来源

随着铸坯应用范围的扩大和质量改进的需要,人们越来越关注夹杂物的特性和危害。连铸夹杂物具有两个突出的特征。首先,它的来源很复杂。钢液有很多机会与耐火材料(中间包衬、塞棒、水口、浸入式水口、中间包和中间包中的结晶器保护渣)接触,钢液大面积暴露在空气中并严重氧化,使钢的清洁度下降;其次,中间包钢液中夹杂物的上浮和分离非常困难。

根据钢厂的实际情况和工艺,通过工业试验开发了几种关键的熔炼、精炼和连铸生产工艺,并得出了以下结论:

(1) 在熔炼过程中必须控制出钢碳含量以控制出钢氧含量,还必须增加炉渣的脱磷能力。

(2) 应选择低碱度、低氧化铁(FeO)的炉渣进行精炼。为使钢液中的铝含量在 0.025%~4.04% 之间,应增加作为沉淀脱氧剂的铝的用量。在 LF 精炼的加热过程中应采用流速为 180L/kg 的氩气搅拌。

5.2.1.2 不同连铸机的夹杂物含量

对于不同的连铸机,铸坯的夹杂物量明显不同。不同铸机每千克铸坯的夹杂物含量为:

立式:0.04mg/kg;立弯式:0.46mg/kg;弧形:1.75mg/kg;水平式:1.35mg/kg。

5.2.2 减少夹杂物的措施

前已述及,温度太低可能堵塞水口而导致铸造困难并导致生产不纯净的铸坯。在脱氧后的钢包处理过程中使用钙基和镁基化合物将氧化铝夹杂物转化为液态铝酸钙,减少了水口堵塞。

提高铸坯纯净度的其他有效方式如下:(1) BOF 无渣出钢。(2) 精炼技术。(3) 无氧化连铸技术。(4) 中间包冶金。(5) 高质量的耐火材料。(6) 电磁搅拌保护钢包注流装置。

5.3 表面裂纹及其控制

常见的表面裂纹如图 5-1 所示。它包括表面横向角裂纹、纵向角裂纹、横向表面裂纹、纵向表面裂纹、星形裂纹、振痕、气孔和大夹杂物等。

5.3.1 板坯表面缺陷及其控制

表面缺陷是严重的质量问题,可能需要精整后合格甚至报废。

(1) 纵向角部裂纹。板坯中很少出现纵向角裂纹。它们的形成可能是由于结晶器对坯壳的支撑不足。结晶器的锥度、表面状况和结晶器冷却水流量可能有重要影响。例如，在美国钢铁公司的得克萨斯工厂，当1200mm长的结晶器被更短的900mm结晶器取代时，缺陷消除了。

(2) 中心纵向裂纹。通常认为，板坯的中心纵向开裂发源于结晶器，并且裂纹的严重性受结晶器正下方的二次冷却和板坯支撑系统的影响。

(3) 星形裂纹。星形裂纹通常与处铜的存在有关，由结晶器和坯壳之间异常的摩擦引起。应注意结晶器表面质量和几何形状（磨损和变形）、润滑（使用具有较高流动性的保护渣）、结晶器壁的温度（一次冷却效率）和结晶器-坯壳之间的接触时间（振动模式）。

低拉坯速度（<0.7m/min）增强了这种缺陷发生的可能。避免星形裂纹最有效的方法是用铬进行结晶器板电镀或用镍进行爆炸电镀；然而，后者很少用于板坯连铸结晶器。

(4) 横向裂纹和振痕。在某些情况下，铜脆产生横向裂纹而不是星形裂纹。

另外，在铸坯被顶弯前，要求铸坯表面温度在低延展性的温度范围之外，以避免横向开裂。

横向裂纹可能出现在表面中心，但它们通常是角裂纹，通常与振痕有关。这些裂纹在弯曲和矫直时打开，因此铸坯的曲率半径影响裂纹产生频率。

这种类型的裂纹发生在含铝量超过0.020%的钢中，并且与AlN沉淀的脆化有关。铌的存在增强了裂纹敏感性。这种缺陷在高强度钢中占主导地位。

(5) 化学成分。钢的化学成分对纵向裂纹的发生有重要影响：碳含量约0.12%、低Mn/S比（约20）、高硫含量（>0.025%）、高锰含量（>1%）、铝含量低于0.004%和铌的存在使钢对裂纹敏感。钢的化学成分对裂纹敏感性的影响可能是由于固相线附近的脆性。临界温度范围因合金元素如硫的存在而扩大，在该临界温度范围内发生包晶转变增加了应变并增加了凝固坯壳的裂纹敏感性。

(6) 横向表面裂纹。连铸中的横向裂纹是由纵向拉伸应变引起的。这种应变在喷淋室中或通过矫直产生。如果表面温度太低，后者可导致裂纹形成，特别是对于低延展性的低温区（700~900℃）范围内。

为了评估横向开裂的风险，引入了两个简单的指标。它们不是明确的标准，因为它们不能明确地预测裂纹产生，也没有临界值可用（比如，如果有临界值，屈服应力就可以用来作为标准化指标）。然而，指标值随着横向裂纹风险的增加而增加，因此可以在不同情况下对板坯的机械性能进行定性分析。结果表明，表面横向裂纹一般出现在角部。Brima-combe和Mintz的研究结论是基于纵向应力和纵向应变率指标的简单直观定义，这些指标已被证实是横向裂纹形成的关键因素。此外，这两个指标也涉及了连铸过程中的不利因素，即在 $T_1 = 700$℃ 和 $T_2 = 900$℃ 之间的温度范围内铸坯缺乏延展性。它们建立在以下基础上：

1) 如果铸坯力学状态有利于扩大横向裂纹，则指数为正，在其他情况下等于零。
2) 当材料变形的可能性很小且没有裂纹时，即当材料的延展性低时，指数是重要的。
3) 指数越高，裂纹风险越高。

避免板坯表面裂纹的方法：裂纹萌发可追溯到结晶器条件和润滑，但裂纹扩展主要取决于二次冷却条件。为避免裂纹，建议在表面温度高于900℃或低于700℃的情况下弯曲或矫直板坯。然而，铸坯的冷却模式对裂纹的发生具有重要影响；有人建议顶部喷淋区的冷却强度应限制在 100℃/m 以下。

5.3.2　大方坯和小方坯的表面缺陷

在进一步加工之前，必须彻底检查大方坯和小方坯的表面质量。

(1) 气孔。气孔是由早期凝固过程中的气体引起的。必须控制钢液中的氢气，并通过真空脱气降低氢含量至 3ppm 以下。钢中的氮含量取决于实际操作。一氧化碳可能在气孔形成中起主要作用。氢和氮含量控制关乎能否充分脱氧。

(2) 气泡。最大限度减小气泡形成概率的工艺参数是高连铸温度、高拉坯速度、保护渣脱水、浸入式水口和高铝含量。

由夹杂物引起的表面缺陷比气孔更频繁、更严重，并且控制它们的形成代价更高。

(3) 钢种成分。表面缺陷的发生取决于钢的成分。硅锰钢具有良好的表现，而含有 0.2% 碳的铬锰钢中，氧化物夹杂很常见。通常，铝添加量超过 0.020% 会显著增加产生夹杂物的可能性，除非脱除氧气。

(4) 纵向裂纹。纵向裂纹的危害非常大，可能导致小方坯和大方坯报废。钢对表面裂纹的敏感性取决于其化学成分，硫含量高于 0.020% 的钢对裂纹特别敏感。

结晶器几何形状（锥度、形状、表面质量）在表面裂纹形成中起主导作用。

浸入式水口和结晶器中的保护渣层可以从根本上改善表面质量。通常，该技术可用于更大的 150mm 的断面尺寸。在某些情况下，由于结晶器夹渣而引发表面夹杂物的问题。

在连铸过程中，为了生产没有表面缺陷的铸坯，控制初始钢液凝固以推断出波状表面的振痕是非常重要的。根据磁流体动力学理论，利用磁场的约束，使弯月面与结晶器壁"软接触"可以产生没有表面缺陷的铸坯。在连铸工艺中，弯月面形状的确定不仅是磁约束计算的关键因素，也是成功实现连铸目标的关键。

板坯连铸工艺由于工艺短、投资少、能耗低和生产效率高，受到了广泛关注。中薄板连铸机效果介于传统板坯连铸机之间，板坯质量较高。研究人员分析了鞍山钢铁公司中厚板的质量，调查了钢包渣、LF 精炼工艺、喂丝工艺、结晶器中钢液流动和钢的连铸工艺的特性。在实际过生产过程中的工艺优化表明，板坯质量明显提高。可以得出以下结论：

(1) 添加助熔剂与石灰后，顶渣的碱度变高，（MnO+FeO）含量降低，从而防止回硫，有利于 LF 精炼。

(2) 钢包温度、化学成分均匀且脱硫后，钢液质量得到显著提高，可满足中薄板连铸的要求。

(3) 通过增加浸入式水口底孔的钢液流速，降低侧孔向上循环的钢流强度，明显提高了结晶器铜板的使用寿命。结晶器液面变得更稳定，并且在窄面附近的钢的湍流变弱。单个结晶器浇注的钢液量从 4 万吨增加到 6 万吨。此外，结晶器保护渣的熔化特性和流动性得到显著改善。

(4) 中薄板坯连铸的钢液经脱氧、LF 精炼、喂丝、中间包冶金等不断清洁，板坯中夹杂物总量可降至 62ppm。

(5) 在钢的生产中，通过降低拉坯速度、优化二次冷却、使用低导热率的中碳保护渣和底孔 $\phi 35 \times 38mm$ 的浸入式水口，可以减少板坯缺陷。板坯宽面上的凹陷和横向裂纹的发生从每米 2 个减少到每米 1 个，板坯合格率达到 98%。

5.4 内部缺陷及控制

内部缺陷如图 5-2 所示。内部缺陷包括内部角裂纹、中心线裂纹、星形裂纹和斜裂纹、疏松、收缩、皮下偏析带和非金属夹杂物等。

5.4.1 板坯内部缺陷

(1) 皮下偏析带。中径裂纹被称为径向条纹或鬼线。这些裂纹通常充满富含溶质的钢液。酸蚀后低倍宏观观察，它们是连铸板坯的柱状晶粒结构之间的薄带状晶间条纹，或者是紧密排列的相互连接的条纹。

这些裂纹可能与机械变形有关，会引起沿凝固前沿的拉伸应变。可以区分两种主要类型的偏析：

1) 平行、细条纹的偏析带，垂直于窄面延伸，似乎与铸坯缺乏支撑有关。

2) 平行条纹或分散的偏析带，垂直于宽面生长。由于与带液芯弯曲、矫直或轧制导致轧辊弯曲、未对准或卡死。

通常认为，优质钢不应该带液芯矫直。然而，Salzgitter AG 的经验表明，带液芯矫直对铸坯内部质量没有不利影响，前提是它不会因拉坯而与额外的拉力相结合。

径向条纹的出现频率与钢的化学成分密切相关。

(2) 中心线裂纹（中心线偏析）。如果矫直或顶弯带液芯的铸坯，则可能形成中心线偏析或内部裂纹。

中心线裂纹似乎是由于在凝固点附近发生的鼓肚或轧辊未对中造成的。该缺陷与铸机的几何形状和形状稳定性有关。中心线裂纹的产生原因如下：

1) 可能是由于熔池末端附近的轧辊未对中造成的。

2) 当熔池由于铸机某些结构不稳定而延伸到拉坯部分时，可能会出现这种情况。在这种情况下，偏析还取决于拉坯辊的设定液压力。

在厚度较小的板坯中，中心线裂纹的发生更为频繁，且随着拉坯速度升高而增加。中心线裂纹似乎与连铸温度无关，但增加冷却强度，在下部的冷却区域，中心线裂纹发生频率降低。

偏析显著损害用于管线或焊接的板坯的质量。

5.4.2 大方坯和小方坯的内部缺陷

通过宏观酸蚀或硫印法可以观察优质钢的内部质量。

内部裂纹可能发生在皮下、对角线或铸坯中心。它们与铸坯的结构及其相关的枝晶偏析有关。裂纹中经常充满富集的钢液，但在某些情况下，裂纹是开放（中心裂纹）。

钢的化学成分对其有很大影响。

高内部裂纹率的条件是：高连铸温度、高拉坯速度、过度的二次冷却、支撑辊和夹送辊的未对中以及过大的拉坯压力。

在连铸对裂纹敏感的钢种时，必须充分控制二次冷却。

(1) 中心孔隙度（疏松）。方形或圆形断面的小方坯和大方坯通常具有中心疏松。在实际生产中，可以识别出两种类型的缺陷。

(2) 收缩。"云状物"：这些是分散在中心区域的众多微空洞。当铸坯以大于6∶1的压下率轧制时，这种类型的缺陷焊合，并且在钢的最终使用中不存在任何不利影响。

(3) 气孔。"孔"：这是一个直径可变的单个中心孔洞。它的形成源于凝固桥和完整的枝晶。如果在加热期间没有被氧化，它会以约10∶1的速率消失。

树枝状结构和高拉坯速度有利于这种缺陷发生。

(4) 皮下气泡。目前为止，唯一的限制因素似乎是易切削钢、含铅钢和用于滚珠轴承的1%C-Cr钢。然而，中碳易切削钢以及含铅钢通常是连铸的。后一钢种表明内在表面质量在进一步加工之前一定要经过调节。一般认为这些钢种只能由立式连铸机连铸。

由于内部裂纹和碳化物偏析的问题，1%C-Cr钢被认为是较难连铸的钢种之一。

5.4.3 减少内部缺陷的途径

(1) 电磁场可用于在钢液中产生强烈的循环钢液流，已经开发了估算流动的理论方法。

在小方坯中，旋转搅拌可以引发熔池中钢液旋转；在板坯中，线性搅拌用于推动钢液沿着水平轴或垂直轴移动。

在小方坯连铸中，具体技术可以不同，但通常，电磁搅拌器是水冷环状结构，感应线圈缠绕搅拌器框架，其位置取决于铸坯尺寸和拉坯速度。

电磁搅拌促进树枝状结构早期向等轴晶结构转变，在不影响连铸条件的情况下，有效降低或抑制了轴向孔隙率。

法国钢厂用法国钢铁研究院开发的电磁搅拌方法改善铸坯组织，其中四流大方坯连铸机在结晶器下方配备搅拌器，铸坯缺陷减少，汽车锻件的疲劳性能得到改善。

在铸坯定尺切割被氧化之前，在线轧制会降低中心孔隙率，电磁搅拌可以补偿高的浇注温度。

对于板坯连铸，最近才应用电磁搅拌。

电磁搅拌的目的是沿着凝固线破坏枝晶并减小柱状晶区的宽度。但由于板坯的宽度，效果比大方坯或小方坯差。

(2) 有利于柱状晶结构形成的条件使偏析问题更严重，即高连铸温度，钢的成分以0.3%，0.1%和0.6%碳含量的逐渐恶化。大断面铸坯和方坯的偏析相对较少。

在板坯连铸中，对未完全凝固的大方坯和小方坯中施加机械应力会产生大的偏析。如果中间包中的钢液温度较低，则可以实现较少的偏析。

(3) 洁净度是大方坯和小方坯连铸最难解决的问题。

对质量要求严格的钢种，需要大量的努力提高洁净度。只有严格控制脱氧操作和连铸条件，才能达到与传统钢种相同的洁净度水平。特别是对于优质钢，钢包和中间包之间的保护浇注、结晶器保护渣覆盖和中间包气氛控制等手段均已采用。

在小方坯连铸（<150mm）中很难实现封闭浇注，其中唯一的注流保护方式是气体保护。而且，中间包钢液的铝含量必须低，以防止小水口堵塞，并且必须将过量的铝供给到结晶器中，否则不利于提高洁净度。

(4) 凝固组织、铸坯支撑和液芯长度三个因素对偏析有很大的影响。

1) 严重的偏析与板坯鼓肚有关。优质钢板坯连铸机在安装时的公差十分严苛，特别是辊的对中、间距和变形。铸坯支撑的效果取决于辊间距和直径。

2)偏析减轻通常与中心等轴晶区的有关。为了减少偏析,等轴晶区域应至少宽30mm。这种中心等轴晶组织取决于化学成分、过热度以及厚度。

在连铸过程中,过热度是影响最终凝固结构的主要操作参数。在大多数情况下,中间包的目标是15℃的过热度,99%的情况下过热度为±12℃之间。在连铸优质钢(高强度钢)时,必须严格控制中间包温度,以便将所需的过热度降低到10℃以下。

产生中心等轴晶结构的另一种方法是电磁搅拌。

1)一般来说,提高拉坯速度加剧偏析。这是较薄的板坯(160mm厚)始终比厚板坯偏析更严重的主要原因。在较低的拉坯速度下,较厚的凝固坯壳在连铸机中的任何给定水平下抗鼓肚能力更强。

2)增强凝固坯壳并减少偏析的另一种方法是增强铸机下部的冷却来使坯壳更冷,一直延伸到最终的凝固点。

在连铸的小方坯和大方坯中,铸坯中心的正偏析通常与完全树枝状组织相关。危害特别大的偏析形式是V形偏析,一般认为是由凝固搭桥。这种缺陷对于冷镦应用是不利的。

5.5 铸坯形状缺陷

铸坯的形状缺陷不仅是物理形态问题,还与裂纹和漏钢等其他问题有关。形状缺陷是由于铸坯表面的不均匀冷却或者铸坯支撑不充分造成的。

5.5.1 鼓肚变形

5.5.1.1 概述

板坯基本都会发生鼓肚变形,它会影响板坯质量。

如图5-3所示,在轧辊支撑力和钢液静压力的共同作用下,坯壳凸起,导致固液界面不能维持一个平面。Kusano等人发现在连铸方向上的固液界面的运动轨迹是由界面的周期性弯曲变形形成的正弦波图形。界面的这种变化对柱状晶区中的溶质分布有影响,也影响内部裂纹的形成。

当固-液界面发生周期性弯曲变形时,两相区域中树枝晶间距的变化如图5-4所示。当固液界面为平面时,两相区域的枝晶如图5-4(a)所示。当固-液界面弯曲变形时,树枝晶间距将在连铸过程中发生变化:在正弦波的波峰处间距最大,如图5-4(b)所示;在正弦波的波谷处间距最小,如图5-4(c)所示。如上所述,固液界面的弯曲变形改变了温度场,沿拉伸方向的温度梯度为二次枝晶的生长提供了条件。当一次晶间距足够大时,将产生二次枝晶。枝晶间距明显增大,二次枝晶在凝固过程的中期和末期快速生长。因此,当界面到达正弦波的波谷时,二次枝晶可能会重叠,如图5-4(d)所示。

如果初生枝晶和次生枝晶都发达,枝晶底部的封闭区域将在波谷形成。如果封闭区域仍然存在于波峰处并且封闭区域的钢液在凝固期间没有被填充,则封闭区域将变成小孔。如果孔隙在随后的变形中没有焊合在一起,它将成为裂纹源。

虽然坯壳相对较薄,但主要的树枝晶间距非常小,并且二次枝晶未发育完全。因此,在中心而不是在连铸板的表面上观察到裂纹源,因为枝晶不重叠。

5.5.1.2 连铸中板坯的鼓肚程度

利用材料力学塑性变形和连铸蠕变变形理论,得到了板坯鼓肚的计算公式。考虑了钢液的计算公式和静压增量的影响。因此,建立了一个新的数学模型。数值计算表明,适当的二冷水量将减小鼓肚的尺寸;拉坯速度越大,鼓肚的尺寸越大,拉坯速度在0.1m/min变化,相应的鼓肚尺寸变化范围为6%~15%;轧辊间距的影响更加显著,辊距的变化为10mm,相应的鼓肚尺寸变化范围为10.5%~21%。

5.5.1.3 减少鼓肚的方法

(1) 降低连铸机的高度。
(2) 降低二次冷却区的辊距。
(3) 辊距应从连铸机上部开始逐渐增大。
(4) 支撑辊装配严格对中,增加二次冷却区的冷却强度。
(5) 防止支撑辊变形,板坯支撑辊应选择多辊。

5.5.2 小方坯脱方

5.5.2.1 概述

小方坯的脱方(偏离正方形)也是连铸过程中的缺陷。当横截面对角线不等时,横截面变为菱形,如图5-5所示。

菱形变形的量表示为 $a-b$ (mm) 或 R (%)。

$$R = \frac{a-b}{\frac{a+b}{2}} \times 100\% \quad \text{或} \quad R = \frac{a-b}{b} \times 100\%$$

当R在某种程度上很大时,它会导致两个钝角的产生斜裂纹。引起对角裂纹的R的程度与钢的热延展性有关。例如,低碳钢(碳含量为0.08%~0.12%)中,$R>3\%$时会产生对角裂纹。这意味着,低碳钢铸坯易于发生菱变,但当$R<3\%$时,它不会引起对角裂纹。1%C-1.5%Cr小方坯中,有斜裂纹,甚至脱方非常低时,斜裂纹和脱方同时出现。铁素体不锈钢热延展性较低,比奥氏体不锈钢更容易产生裂纹。在裂纹处偏析会富集,这对钢的热处理性能产生不利影响。

5.5.2.2 脱方的原因

在冷却开始时,小方坯四角的冷却不均匀引起了脱方。强烈冷却的两个对角将成为锐角;对应的角将成为钝角。

5.5.2.3 降低脱方的方法

(1) 使用两个锥形结晶器和抛物线型结晶器。
(2) 结晶器的狭窄水缝和高刚度。

(3) 降低结晶器液面波动。
(4) 保持对弧的精度。
(5) 保证二次冷却区的冷却。
(6) 钢液成分控制。

5.5.3 圆坯的变形

5.5.3.1 引起圆坯变形的原因

圆坯可以变成椭圆形或不规则的多边形。横截面直径越大，圆坯变形越严重。引起圆坯变形的原因如下：
(1) 结晶器内部形状发生变形。
(2) 二次冷却区冷却不均匀。
(3) 连铸机下部的对弧准精度低。
(4) 拉矫辊夹紧力不合适，不能轻压下。

5.5.3.2 减少圆坯变形的相应措施

(1) 及时更换变形的结晶器。
(2) 确保连铸机对弧准精度的技术。
(3) 二冷区冷却均匀。
(4) 适当降低拉坯速度。

5.6 漏钢

漏钢是最严重的事故。当发生漏钢时，坯壳开裂，钢液在结晶器下方倾泻而出。漏钢是炼钢过程的一个重要事故，因为：
(1) 停机时间导致产量率减少。
(2) 因修理或更换损坏的设备而成本增加。
(3) 最重要的是，对钢厂操作员构成了重大的安全风险。据报道，传统板坯连铸机的漏钢损失可能接近20万美元。因此，为了实现连铸工艺的稳定操作，必须减少漏钢。

在开浇期间的漏钢称作开浇漏钢，在后续运行时操作期间称作运行漏钢。Dofasco 的经验表明，大约25%的漏钢实际上都属于开浇漏钢。通过降低拉坯速度可以在一定程度上避免开浇时和连铸过程中漏钢，从而在结晶器中提供更多的停留时间以使钢液凝固。为了避免发生漏钢，必须提前检测坯壳的异常凝固，并有足够的准备时间来适当减慢铸机的拉速。

5.6.1 漏钢的原因

(1) 浇注温度（过热度）。如果浇注温度太高，坯壳会很薄。由于结晶器保护渣和钢的静压，坯壳和铜板之间的摩擦力升高，从而引起粘连。
(2) 结晶器保护渣的性能。结晶器保护渣润滑性能差是导致漏钢的主要原因。结晶器保护渣的参数包括熔化温度、熔化速率、结晶温度、凝固温度和黏度等。结晶器保护渣具有良好的润滑性，使连铸温度适宜时连铸工艺顺畅；如果连铸温度过高或过低导致润滑不良，则坯壳与铜板之间的摩擦力升高，可能导致粘连。

（3）钢的成分。碳不仅是钢中的基本元素，也是结晶组织中最有效的元素。当钢的碳含量为约 0.12% 时，热通量最小。坯壳相变产生体积收缩，坯壳与结晶器之间形成气隙，此时热通量最小，坯壳薄而不均匀。由于钢的静压，结晶器出口拉坯时，可能发生裂纹。

（4）拉坯速度。一般而言，拉坯速度、结晶器保护渣的供应速率和结晶器铜板的冷却水是稳定的。当拉坯速度突然改变时，结晶器铜板的温度波动和结晶器保护渣的不连续供应将导致坯壳直接接触铜板，可能导致漏钢。

5.6.2 防止漏钢的措施

（1）选择性能优异的结晶器保护渣。
（2）确保连铸过程操作稳定。
（3）结晶器精度高。
（4）结晶器振动良好。
（5）实时监测预测系统。

5.6.3 多变量 PCA 防止漏钢

已经提出了许多方法来预测连铸中的漏钢，包括：

（1）模式匹配法（Yamamoto, Kiriu, Tsuneoka&Sudo, 1985; Nakamura, Kodaira, &Higuchi, 1996）。

（2）MPCA（Multiway Principal Component Analysis 主成分分析）法。粘连检测法是典型的方法，依据以往漏钢发生之前结晶器的温度图像（由结晶器周围的热电偶测量）。MSPC 法也已应用于漏钢预测（Vaculik 等，2001），其中测量选定的过程来建立 PCA 模型以模拟连铸工艺的正常操作；然后由模型计算某些统计数据以检测包括漏钢在内的异常操作。

遗憾的是，这两种方法都不能用于检测和防止由于某些技术的限制而导致的漏钢。因此，需要提供足够的预报时间的实时在线监测操作并检测即将发生的漏钢，以便采取适当的相应措施来防止漏钢。一种可能的应对措施是改变拉坯速度的增长率以减慢连铸工艺，并为结晶器中的钢凝固提供更多时间。

5.7 喷水冷却对铸坯质量的影响

除了结晶器-冷却器距离之外，结晶器出口温度和冷却速率是影响凝固前沿位置的最重要因素，并影响前述的一些铸坯表面缺陷。关键要保持结晶器出口温度恰好高于钢液的凝固温度。必须适当增加冷却速率以满足冷却铸坯的需求。提高冷却速率主要通过增加冷却水流速和使用多个冷却装置来完成，以增加被冷却铸坯表面积。在使用结晶器冷却装置的情况下过长的冷却装置等同于使用多个较短的冷却装置的效果。例如，长 10cm 的结晶器冷却装置超过了将铸坯冷却至室温所需的约 1.5cm 的长度。优点是，8.5cm 的过冷段具备自动化的可能。这产生了一个主观的结论，即过冷段可能是增加冷却水量的替代装置，或是减少结晶器-冷却器距离作为提高拉速导致的冷却速率增加的补偿。

铸坯中最严重的缺陷是喷水冷却系统设计不当引起的裂纹。

在结晶器下方，来自凝固坯壳内的钢液的压力可以引起辊间铸坯鼓肚，导致凝固前沿的应变，导致裂纹形成，并且渗入富含溶质的钢液（偏析）。通过精心设置的喷嘴系统来防止这类的裂纹，以产生能够承受铁水静压力的坯壳。这种坯壳的特性与低于1200℃的表面温度有关。

中间裂纹与沿连铸机的铸坯表面被加热有关。再加热是当铸坯从结晶器移动到喷嘴、从喷射区域到下一个喷射区域或从喷嘴区域到辐射冷却区域时，表面散热量突然降低的结果，结果造成表面趋于膨胀并且凝固前沿处产生的拉伸应力引起热裂。中间裂纹的允许再加热量取决于许多因素。其中一个最重要的因素是铸坯组织。具有大量等轴晶组织的钢可以避免裂纹的形成，直到再加热200℃，而较低的表面再加热温度产生的柱状晶结构更有利于裂纹产生。

一些研究人员致力于开发基于有限差分法的二维传热模型，以计算沿铸机方向的铸坯温度和坯壳轮廓。人工智能启发式搜索程序与数值模型相互配合，以完善真实连铸机喷淋区域冷却条件，以生产高质量的铸坯。

6 连续铸钢新技术

自 20 世纪 70 年代以来，连续铸钢取得了重大的新成就：如结晶器液面控制和带热电偶的结晶器、电磁搅拌（EMS）、中间包冶金、温度和钢液流动的数字化、结晶器中的电磁制动（EMBr）、薄板坯连铸、液芯压下（LCR）、动态轻压下、高速连铸、结晶器内流动控制（MFC）、动态喷嘴控制、近终形连铸、热装轧制、带钢直接浇注和新兴的 3D 在线控制。这些技术创新提高了生产效率和铸坯质量，有助于提高资源利用率。在本章中，将分别介绍其中的一些创新成果。

6.1 中间包冶金

在连铸机中，钢液从钢包转移到结晶器涉及两个步骤。首先将钢液从钢包连续或半连续地浇注到中间包，中间包又通过水口以连续流动的方式将钢液分配到各个结晶器中。钢包水口进入中间包的金属流动由塞棒机构或液压、电控滑动水口系统控制。（由于更强的控制能力和可靠性，后一种系统正在迅速普及。）

因此，中间包应从钢包中接收钢液并将其分配到各个注流上，同时尽可能减少热量损失、防止钢液二次氧化和夹杂物通过。

为了高效生产高质量产品，在炼钢过程中需要有效的钢液清洁技术。夹杂物分离和除渣对钢液的洁净度很重要。因此，采用了以下技术，但会在中间包中引入杂质：

(1) 使用大容量中间包以获得足够长的钢液停留时间以使夹杂物上浮；
(2) 改善坝体，尽量控制钢液流动；
(3) 通过加热中间包中的钢液促进夹杂物分离；
(4) 中间包钢液吹氩；
(5) 开发一种在钢包更换期间可以从两个钢包供给钢液的工艺；
(6) 中间包表面覆盖以防止空气氧化；
(7) 开发了钢液通过电磁力水平旋转的离心流动中间包。

以下主要介绍中间包加热技术、中间包流场控制、中间包保护渣和离心流中间包。

6.1.1 中间包加热技术

中间包最初盛满钢液时，熔池温度降低到所需温度以下。由于温降也会引起麻烦的水口堵塞问题，因此配备有加热装置的中间包是优选。

高温中间包的主要优点是可以将出钢温度降低 50℃，从而节省炼钢炉耐火材料成本，降低回磷的风险。后者在冶炼高磷铁水时优势更加明显。

加热装置有两种：感应加热和等离子加热。这两种装置都可确保在整个连铸期间将熔池温度波动控制在 5℃ 以内。图 6-1 中所示的是一个实例，等离子加热确保了中间包均匀的熔池温度。加热装置使向中间包中添加少量合金元素来调节熔池成分成为可能。

如果热的中间包在剩余的渣和钢液倒出后被重复修复和使用,可以大幅降低成本。然而,缺点是下一炉会被中间包中残留的渣污染;通过使用一些具有高碱度和低黏度的助熔剂克服了这一问题。图 6-2 说明了如何重复使用热中间包。

此外,数学模型研究表明,中间包中熔体的传热和流体流动特性不能单独处理,而是必须与钢包的传热特性相结合,以实现连铸过程的真实模拟。

6.1.2 中间包流动控制

连铸操作中的中间包是钢包、转炉/电炉和结晶器之间的桥梁,具备连续操作的特性。传统上用作钢液的储存和分配器,但是现在,它的作用已经大大扩展到提供所需洁净度和成分的钢液。因此,在连续铸钢中,流动系统的任务是将钢液以所需的流速从钢包输送到结晶器中,并确保将钢液输送到既不太冷湍流也不明显的弯月面。此外,钢液流应尽量减少暴露在空气中,避免夹渣或夹带其他异物,有助于夹杂物上浮到渣层,并促进均匀凝固。实现这些有些矛盾的任务需要细致的工艺优化。

中间包钢液流动应设计为增加钢液停留时间、防止"短路"和促进夹杂物的去除。中间包钢液流动由中间包的几何形状、液面高度、水口设计和流量控制装置控制,例如冲击垫、堰、坝、挡板和过滤器。

中间包冲击垫是一种廉价的流量控制装置,可抑制湍流并防止钢液冲击中间包底造成腐蚀。进入中间包的注流动量是扩散的,钢液的自然浮力避免钢液流短路,特别是在开浇时。TURBOSTOP 浇注垫与堰和坝一起改善了钢的洁净度,特别是在钢包更换期间。在卢肯斯钢厂,全氧含量从 26ppm(带拱顶型防冲击垫)减少到 22ppm(带毂盖形防冲击垫)。在浦项钢铁公司,在坝上开 77 个孔来改善钢的洁净度,起到过滤器的作用。在 Dofasco 的 2 号炼钢车间,挡板的使用提高了铸坯质量,特别是在换钢包时使温度更加均匀。挡板和中间包表面覆盖一起,使中间包钢液中全氧含量从 39±8ppm 降到 24±5ppm。

6.1.3 中间包保护渣

中间包保护渣必须具备多种功能。首先,它必须满足绝热(以防止过多的热量损失)和化学隔离(以防止空气夹带和再氧化)。例如,在 Imexsa,通过改变中间包保护渣(具有较低的 SiO_2 含量),钢液从钢包到结晶器,氮气吸收从 16ppm 降至 5ppm。

其次,在理想情况下,保护渣还应吸收夹杂物,以提供额外的精炼效果。一种常见的中间包保护渣是碳化稻壳,它价格低廉,不易结晶,是一种良好的绝热体。然而,稻壳中的二氧化硅含量很高(SiO_2 80%),SiO_2 可以形成夹杂物。稻壳灰尘含量和碳含量高(C 10%),可能会污染超低碳钢。

基础保护渣(CaO-Al_2O_3-SiO_2 基)理论上稻壳效果更好,中间包钢液中全氧更低:T.O 从 25~50ppm 下降到 19~35ppm,实际上保护渣碱度从 0.83 增加到 11。在一些炼钢车间,与初始保护渣(CaO 3%,Al_2O_3 10%~15%,MgO 3%,SiO_2 65%~75%,Fe_2O_3 2%~3%)相比,基础中间包保护渣(CaO 40%,Al_2O_3 24%,MgO 18%,SiO_2 5%,Fe_2O_3 0.5%,C 8%)与挡板同时使用可显著降低全氧含量。在钢包过渡期间 T.O 从 41ppm 降至 21ppm,在稳态浇注期间从 39ppm 降至 19ppm。

然而,有些例外如图 6-3 所示,发现使用稻壳和较高碱度保护渣(SiO_2 25.0%,Al_2O_3

10.0%，CaO 59.5%，MgO 3.5%）后，T.O 没有改善。这可能是因为碱性保护渣仍然含有太多的二氧化硅。更可能的原因是，由于较快的熔化速率和较高的结晶温度，基础保护渣效果大打折扣，因为它很容易在表面形成硬壳。而且，碱性保护渣黏度通常较低，因此更容易夹渣。可以使用双层保护渣避免这些问题：底部是低熔点基础保护渣吸收夹杂物，而顶部的稻壳提供绝热，从而 T.O 从 22.4ppm 降低到 16.4ppm。

6.1.4 离心流中间包

人们设计了一种新的工艺来促进中间包中钢液夹杂物的分离。该过程利用电磁力将圆柱形中间包中的钢液旋转。由旋转流动产生的离心力促进了夹杂物与钢液的分离。这种中间包被称为离心流中间包（CF 中间包）。

CF 中间包如图 6-4 所示。从圆柱形中间包外部施加移动电磁场。钢液水平旋转，并且由于夹杂物相对于钢液的密度较低，向心力作用在夹杂物上，促进钢液与夹杂物分离。洁净的钢液从中间包底角流入结晶器。

CF 中间包中夹杂物分离和除渣的机制如下：
（1）通过向心力聚集夹杂物；
（2）促进夹杂物碰撞和长大；
（3）通过旋转流改善停留时间分布。

千叶公司生产设备的测试结果表明，CF 中间包具有优异的脱氧性能，脱氧速率约为 $0.17 \sim 0.25 min^{-1}$。由旋转流引起的向心力和大量湍流可加速夹杂物分离。此外，改善了中间包中钢液的停留时间分布，并促进了钢包更换过程中的除渣。该工艺已成功用于高品质不锈钢板的规模化生产。

6.2 结晶器液面控制

结晶器液面控制如图 6-5 所示。

连铸机控制中最重要的是确保铸坯顺利拉出的同时使得结晶器中的液面保持恒定（在几厘米内）。有两种手段：

（1）中间包称重，从钢包到中间包的进料速率自动变化，以保持中间包总重量不变。以这种方式，中间包进料的速率是恒定的；
（2）控制铸坯拉出速度，保持结晶器液面大致恒定。

在连铸早期，通过操作者观察并且相应地调节中间包塞棒，使连铸机中的液面保持恒定。现在可以使用测量仪器自动调节液面高度。表 6-1 列出了检测液面的几种方法。下面详细描述其中的两种，即伽马射线（放射性）和红外线方法。

从图中可以理解该操作。研发红外设备是为了避免使用强放射性同位素。探测器可以观测到金属层与结晶器壁的连接处。随着液面在观测范围内上升，单个光电管接收更多辐射，输出信号更强。制定了特殊规定以应对观测的中断。光电管装置接收红外辐射并向控制单元提供电信号，控制单元又连接到操作人员和连铸驱动装置。操作人员可以选择自动或手动控制，并从信号灯接收操作塞棒的信号。从钢液发出的辐射通过开槽的掩膜准直，然后通过柱面透镜聚焦到光电探测器上。光线经过过滤，消除了波长 1 mm 以下的辐射，从而减少环境和油火的干扰。

整个系统有两个探测器和两股光束,通常布置为观察钢流的各个侧面。通过改变掩膜中的槽间距,可以调整光电管两个探测区之间的间隔。

每个通道都安装了三个光电探测器:第一个使用光束测量液面;第二个接收光束并启用温度漂移补偿;第三个通过位于正常金属层之上的小区域以及主光束和金属流之间的槽,目的是检测金属流是否在中心位置停留并且存在干扰主光束的危险。两个主光束之间的平衡和钢流探测器的阈值可以通过安装在装置背面的小电位计进行调节。

在温度补偿之后,将每个通道检测到的电信号反馈到选择最大信号的简单电路。因此,该单元始终控制两个电信号中的较大者。如果光电探测器观测到铸流向检测光束移动,注流将阻挡信号,并且装置切换到另一个通道控制。还有一个特点是,如果两个通道都被阻挡,例如通过扇形金属流,则该单元切换到存储器,相当于快速检测液面,并防止连铸突然失控。随着存储量衰减,钢液面逐渐下降,使操作人员有足够的时间进行干预。

如果转换时拉坯速度大幅提升,则通过防止自动操作,该装置可以实现从手动控制到自动控制的平稳过渡。从自动更改为手动时,它不提供无扰动传输。当存在电缆故障时,还可以防止变为自动控制。

控制系统接收选择的电信号,并遵循比例和积分作用。直接向拉坯驱动装置输出电压信号。驱动装置产生与该电压信号成比例的拉坯速度。

6.3 板坯热装和直接轧制

尽管将铸坯热装入精轧机加热炉中的实际操作不一定提高生产率,但是由于节约了燃料而受到广泛关注。

在连续铸钢的早期研发中,将铸坯冷却至环境温度,检查缺陷,并在必要时进行精整以去除表面缺陷(这种做法与许多铸锭轧制的铸坯相同)。然后将铸坯再加热并在精轧机中进一步加工,这是高耗能和劳动密集型生产方式。通过将热坯送入精轧机中,铸坯的显热得到充分利用,同时显著节约能源。这种做法可以避免完全再加热或中间再加热。但是,它要求连铸机和精加工区域之间的紧密协调。它还要求铸坯具有优异的表面质量,因为在线热检和铸坯精整尚未完全开发。主要工艺如图6-6所示。

6.4 凝固末期轻压下

众所周知,连铸坯的主要缺陷是中心线收缩和偏析。已知铸坯中心线偏析是由凝固的末期的枝晶间钢液流动形成的。导致枝晶间钢液流动的主要因素是凝固收缩和凝固壳的鼓肚。随着对优质钢的需求不断增加,即使半径小到几毫米的偏析消除也已成为连铸工程师的主要目标之一。

凝固末期轻压下对铸坯内部质量有明显的影响。近凝固端的轻压下技术起源于20世纪70年代末至80年代初期。轻压下技术于1976年在日本新日铁首先成功投入使用。1986年,福山钢厂4号板坯连铸机安装了改进的压下装置。目前,它在全球范围内应用于板坯和大方坯连铸机,被认为是进一步改善大方坯宏观偏析的最佳选择。作为提高铸坯质量和开发高附加值铸坯的重要措施,轻压下技术已成为现代连铸机的组成部分。提出了许多基于热应力模型和收缩准则函数的定量模型,能够更准确地估计压下的效果和位置,更加科学地实施该技术。

6.4.1 轻压下的效果

(1) 宏观组织的改善。由于对铸造过程中凝固收缩的补偿，轻压下可明显改善宏观组织。1 级中心偏析频率、中心孔隙度和疏松均明显下降，1 级及以下偏析频率大幅增加。

压下量越多，结果越好（见图6-7）。对于高碳轨道钢，由于凝固收缩率较大，随着压下量从 3mm 增加到 7mm，宏观组织持续得到改善。甚至压下量达到 7mm，都没有发现内部开裂。轨道钢在正常拉速下，7mm 的压下量是设备的极限，所以应该尽可能地使用这种压下量的高碳钢。

压下区域的位置和长度对中心偏析的影响远大于压下量的影响，因此影响凝固端部位置的铸造参数会显著影响轻压下的冶金结果。铸造参数的变化改变了压下区域长度和最大压下量。例如，在 0.75m/min 的拉速下，5 组机构可以进行轻压下，最大压下量可以达到 7mm，但是对于 0.7m/min 的拉坯速度，只有 4 组机构可以进行轻压下，并且最大压下量减少到 5mm，弱化了轻压下的冶金结果。

(2) 改善成分均匀性。轻压下会影响大方坯的横向和纵向碳分布。图 6-8 说明了轻压下对纵向中心碳偏析的影响。随着轻压下的增加，中心碳偏析指数 C/C_0（C 和 C_0 分别为固体钢和钢液的组成）的波动与没有压下的大方坯相比大大减小。

U71Mn 大方坯，当铸造条件为过热度 = 31℃，v_c = 0.70m/min，轻压下 5mm 时，没有轻压下的大方坯的碳偏析指数分别为 0.93~1.08 和 0.96~1.03。严重的碳偏析可能导致在高碳钢中形成脆性相，例如渗碳体和马氏体。此外，通过检查轨道钢的化学成分分布，Guijun Li 等发现两个随机点之间的成分偏差范围如下：C 0.02%；Mn 0.05%；Si 0.02%；P 0.005%；S 0.005%；并且可以满足 350km/h 的高速铁路的要求。

(3) 机械性能。截至 2004 年 4 月 30 日，攀钢的钢轨和异型坯厂已经浇注了 354500t 重轨。据统计，重轨的机械性能非常好。由于精确的成分控制和通过轻压下获得的偏析较少，拉伸强度和韧性范围在一个狭窄的区域内，例如，对于 PD3，拉伸强度为 1030~1060MPa（平均值为 1045MPa），伸长率为 10.5%~12.0%（平均 11.4%）。在标准落锤试验中，未断裂，试样的挠度在 37~44mm（平均 39mm）内；所有产品均符合 350km/h 的高速铁路要求。

6.4.2 轻压下需要注意的问题

从上述分析得出，轻压下是消除铸坯中心偏析和中心疏松的有效方式。然而，如果在不合适的时间（此时固相分数不合理）轻压下，则对小方坯的内部质量，即小方坯内部裂纹有不利影响。这些裂纹位于小方坯纵断面上的柱状晶区，垂直于浇注方向并填充有残余钢液。停留时间和固相分数对内部裂纹的影响如图 6-9 所示：随着停留时间从 180s 增加到 270s，小方坯内部裂纹指数从 50mm 减小到 0mm，并且铸坯中心的固相分数从 0.65~0.78 增加到 0.96~1.0。这个结果是合理的，因为随着停留时间的增加，凝固坯壳变厚。此外，在长于 270s 的停留时间，固相分数大于 0.96。在如此大的固相分数下，液相穴尖端处的冷却速率高于铸坯表面上的冷却速率，铸坯中心的体积收缩较大。如果铸坯中心的凝固收缩没有得到补偿，则宏观中心裂纹将通过铸坯中心的切向和轴向拉伸应力形成。

通过使用单对和多对辊进行轻压下试验，研究了压下量和压下时间对中心偏析的影响。研究发现，仅在凝固末期（其中板坯中心的固体部分大于 0.25）进行轻压下时，中心偏析随着压下量的增加而提高。同时，还改善了浇注方向的变化和板坯横向中心偏析的分布。人们认为，由于在宽泛的凝固阶段的轻压下导致中心偏析的恶化可归因于轻压下引起的新的钢液流动。为了改善中心偏析，有必要优化压下时间并抑制由于诸如压下辊弯曲的机械因素导致的不均匀压下。

从上述分析可以看出，应用轻压下，必须同时考虑内部裂纹、宏观裂纹和中心偏析。当固相分数非常小时，通过轻压下容易引起内部裂纹。如果固相分数非常大，则不能有效地消除中心偏析，并且在铸坯中会出现宏观裂纹。

6.5 电磁搅拌技术

虽然连铸可以提供更高的铸坯产量、更优质的铸坯和更高的总体生产率，但仍然存在一些相当大的质量缺陷，主要出现在三个区域，即表面、皮下和中心。已经找到了一种成功的解决方案，即电磁搅拌。

6.5.1 电磁搅拌的发展

尽管电磁搅拌技术仅是近年来在连铸工艺中得到应用，但是在大规模连铸出现之前，在凝固过程中搅拌金属以试图更精确地控制工艺的设想已经萌发。

第一个适用于连铸的电磁搅拌装置设计于 1952 年。该装置位于结晶器下方。关于结晶器内搅拌的第一次试验在 Boehler 的 Kapfenberg 进行。当时使用工业频率（50Hz）和低导电率的结晶器材料（铝青铜或钼），以保证穿透到结晶器中的旋转磁场频率为 50Hz。该实验表明铸坯的表面质量得到改善。到 20 世纪 60 年代后期，法国、英国、苏联和美国的小方坯和大方坯连铸机都进行了结晶器内搅拌的研究。此时已经研发了旋转搅拌器和套管搅拌器。这些早期调查证实了铸坯的冶金质量的实质性改善。同样从这项工作中，发现了"白亮带"，并给出了消除"白亮带"的建议，例如在大部分铸坯上使用非常温和的搅拌。

电磁搅拌的第一次成功应用可以追溯到 1973 年由法国钢铁研究院/CEM 在法国 SAFE 的 240mm 2 流大型连铸机上进行。从那时起，电磁搅拌在连铸机上的应用迅速推广到全世界。

日本在此时开始发展电磁搅拌，很快在该领域处于领先地位。到目前为止，他们已经在连铸机上大规模采用电磁搅拌。

首先研发的是用于结晶器下搅拌的各种电磁搅拌装置。这些包括旋转搅拌器、圆柱形线性搅拌器、螺旋搅拌器和水平线性搅拌器。中心线偏析、孔隙度和 V 形偏析的改善已见诸报道。

结晶器内搅拌首先由法国钢铁研究院研发，后来由 BSC 研发。连铸时使用低频，使得行进磁场可以穿透传统的铜结晶器，从而改善了表面和皮下质量。

后来使用了结晶器内旋转搅拌，报告声称，除了在表面和皮下质量方面获得的改善之外，中心质量也得到了改善。

还报道了一些不寻常的结晶器内搅拌器类型，包括用于板坯连铸机的所谓"电磁制动器"、主频结晶器搅拌器和液压驱动的永磁搅拌器。

到 70 年代中期，经过广泛的工业应用，单一电磁搅拌搅拌器的局限性逐渐显现，这表明：要得到令人满意的铸坯质量需要组合搅拌，即在不同位置同时使用几个搅拌器，例如在结晶器内、二冷区和凝固末端同时使用搅拌器。

由于问题的复杂性，板坯连铸机的电磁搅拌的发展落后于小方坯和大方坯连铸机几年，相关工作始于从结晶器下部搅拌，起初使用感应搅拌器和导电搅拌器。

板坯连铸机的结晶器内搅拌由法国钢铁研究院于1978年开始研发，使用线性垂直搅拌器。后来研发了水平线性搅拌器。

自第一台电磁搅拌装置安装在连铸生产线上仅仅几年时间。但是今天，全世界已经安装了数百甚至更多的电磁搅拌器。

6.5.2 电磁搅拌的作用

电磁场可用于在钢液中产生强烈的循环钢液流动模式。已经开发了估算钢液流动的理论方法，在凝固过程中从钢液的电磁搅拌得到的潜在益处受到广泛关注。报道的改进包括：

（1）通过更好的凝固组织改善内部质量（减少偏析，裂纹和孔隙率）；
（2）改进钢液流动模式提高亚表面和内部洁净度；
（3）降低浇注参数（温度和拉坯速度）的临界值；
（4）提高拉坯速度提高生产率。

电磁搅拌使操作者能够在保持拉坯速度恒定的同时提高铸坯的质量，或者在不损害铸坯质量的情况下提高拉坯速度。

法国钢铁研究院开发的改善铸坯铸态组织的电磁搅拌方法用于法国钢厂，其中四流大方坯连铸机在结晶器下方配备有搅拌器，成品的缺陷减少，汽车锻件的疲劳性能得到改善。

美国的测试显示出类似的结果，但这种改进还不足以抵消靠近结晶器安装搅拌器的操作难度。

对于小直径的高碳钢铸坯，日本的研究表明，与未经搅拌的铸坯相比，经电磁搅拌后的铸坯在拉伸过程中产生的断裂要少得多。

在英国，已经使用结晶器中的搅拌器测试了钢液沿着铸坯凝固表面的向上运动，通过电磁搅拌诱导，以寻求改善的表面质量和较低的偏析。

今天，世界上几乎所有优质的方坯/大方坯连铸机都采用这种方法。不锈钢板坯连铸机也是如此。自20世纪80年代以来，他们已经认识到电磁搅拌对等轴晶区或晶粒尺寸控制的好处。电磁搅拌最近才应用于板坯连铸。但板坯和方坯电磁搅拌的目的相同，即沿着凝固线破坏枝晶并减小柱状区的宽度。但由于板坯较宽，要达到相同的效果，比大方坯或小方坯更难实现。

然而，对于在常规厚板坯连铸机上铸造的低碳钢，电磁搅拌的应用仍然受到限制。DC刹车由KSC和ABB在1982年推出。他们在几台弧形结晶器上进行了测试，取得了一些成功，但最初没有突破。因垂直/弯曲仍然是抵消重力的最佳方式，由位于弯月面的两个长线性搅拌器引起的旋转搅拌由Nippon Steel引入，以解决坯壳成形和板坯皮下质量问题。为了改善中心偏析，深入研究了电磁搅拌（EMS）。然而，EMS不能成功应用于板坯连铸机，因为没有偏析的等轴晶的堆积对EMS在液相穴端附近引起的钢液流动模式非常敏感。EMS 应用也仅限于中碳钢，因为其他钢材难以进行晶粒增殖。

经过电磁搅拌的板坯的中心偏析较少，因此焊接性能好。

6.5.3 电磁搅拌器类型

电磁搅拌器最初安装在小方坯铸机上以减少中心偏析。这是通过凝固组织的变化实现的：中心等轴晶区增加，外柱状晶区相应减小。据一项最近的调查报道，有超过 100 台搅拌器在运行，其中 60 多台用于方坯和大方坯机，大约 40 台用于板坯连铸机。有两种基本类型的搅拌器，（旋转式和线性式），可安装在结晶器内或结晶器下方。

(1) 旋转搅拌器。如图 6-10 所示，在安装在铸坯连铸机结晶器中的旋转系统中，由线圈产生的旋转磁场赋予钢液圆周运动。产生的离心力产生良好的坯壳，较轻的相（即夹杂物）朝向中心移动。中心等轴晶区域被扩大，因为旋转流动促进了柱状枝晶尖端的破裂，使其作为中心区域等轴晶体的形核核心。

在铸坯中，旋转区域可以引发熔池中的金属旋转，并且在板坯中，线性区用于沿水平轴或垂直轴移动液态金属。

在小方坯连铸中，具体技术可以不同，但通常，电磁搅拌器是水冷环，具有围绕铸坯的感应线圈，定位位置取决于铸坯尺寸和拉坯速度的高度。法国的 Sherurgie Francise 研究所（法国钢铁研究院）和 Compagnie Electro-Mecanique（CEM）开发的搅拌器，直径约为 0.5m，线圈与底部之间的距离约为 0.5m。根据铸坯的尺寸，结晶器的变化范围为 2~4m，频率为 50Hz，功率通常为每小时 1~5kVA。

(2) 线性电磁搅拌器。利用线性搅拌系统，电磁线圈沿着铸坯的侧面（在结晶器下方）安装，产生垂直的循环模式。通过与旋转搅拌器相似的机理增加中心等轴晶区。夹杂物通常集中在铸坯内弧侧，且分布更均匀。

板坯连铸机电磁搅拌器至少有两种，但它们的细节目前是保密的。在日本，特别是在新日铁，搅拌器由两个直线电机组成，这两个电机位于靠近铸坯的水冷箱中，小的辅助辊压在铸坯上。500kVA 电机提供周期性脉冲以搅动和破碎枝晶，导致它们下沉。

迪林根钢铁公司使用法国钢铁研究院开发的搅拌器进行了测试，该搅拌器由四个线性电动机组成，这两个电动机放置在两对非磁性轧辊内，外径与它们取代的两对传统辊完全相同。

另一种电磁原理应用与板坯连铸机，磁场改变结晶器内某些区域的钢液流动模式并在其他区域内创造钢液流动模式，随后改善内部洁净度和表面质量。这种效果是通过静止磁场中移动的钢流的相互作用实现的。运动的金属流通过磁场产生感应电流，感应电流与稳恒电场一起产生制动力。另外，钢流和电极之间的钢被加速，产生强烈的搅拌。因此，离开耐火水口端的金属流的速度以及穿入钢液的深度减小。在这些条件下，夹杂物浓度降低，并且在结晶器周围发生更均匀的坯壳生长，这减少了表面缺陷的可能性。

因此，在各种搅拌器中，为特定的铸造工艺选择合适的搅拌器并不容易。应该主要考虑冶金目的，在此基础上，Birat 为小方坯和大方坯浇注提供了一些指导原则，如表 6-2 所示。

尽管已经提出并且正在使用各种搅拌器，但是还在开发经济且有效的电磁搅拌装置，还需要做很多电气工程方面的工作。

6.5.4 电磁搅拌的冶金效果

在连铸过程中存在一些相当严重的质量缺陷。铸坯的三个区域具有特定的缺陷，由 H. S. Marr 在表 6-3 中总结。

考虑到表 6-2 中所示的问题，下面解释如何应用电磁搅拌改善铸坯质量。

针孔缺陷是铸坯表面凝固过程中从溶液中散发出的溶解气体产生的小气孔。气孔是被困在铸坯皮下的较大气泡。通过在凝固前沿上的电磁搅拌有效地防止了气泡附着在凝固前沿。

夹渣不仅是耐火材料侵蚀或熔渣进入结晶器时产生，在脱氧产物上浮或钢被大气氧化时也会发生。电磁搅拌可以改进钢液流动模式，钢液流动模式将非金属夹杂物引导到排除位置。特别是在旋转结晶器搅拌的情况下，向心力将非金属夹杂物拉离凝固前沿。非金属夹杂物可被生长的枝晶捕获。通过电磁搅拌产生的强对流、高速度减少夹杂物。这种所谓的"洗涤效果"解释了通过所有类型的结晶器内搅拌获得的清洁表面的事实。

表面和皮下质量只能通过结晶器内搅拌来改善，因为当铸坯从结晶器中拉出时这些区域已经凝固。

在凝固前沿相遇的中心轴或平面上会发生严重质量问题。其中一个主要问题是中心偏析和 V 形偏析的程度加剧。宏观偏析是固-液界面处溶质排出的结果，随后是富溶质钢液沿固体和钢液糊状区中的枝晶内的质量传输。有两种不同的凝固模式，即柱状和等轴状。在柱状凝固模式期间产生糊状区。因此，任何影响从柱状生长向等轴生长早期转变的因素将减少宏观偏析的程度并改善铸坯质量。该领域的几乎所有报告都声称通过使用任一类型的结晶器下搅拌器和结晶器内旋转搅拌器均可增加等轴晶区的尺寸。在应用电磁搅拌之前，通过使用低温度梯度、低过热度和低拉坯速度来控制偏析问题。这种方法不适用于优质钢，且严重限制了工艺灵活性。电磁搅拌的应用放宽了这些操作限制，从而提高了生产率。

中心疏松是另一类中心缺陷，受横截面尺寸变化和凝固收缩量的影响，电磁搅拌产生的流动将消除搭桥，使凝固前沿平滑，促进钢液的有效补给，以适应收缩，从而抑制疏松的形成。

搅拌和未搅拌铸坯机械性能的比较结果显示，拉伸性能、疲劳性能和韧性等改进令人欣慰。

一种相对较新但发展迅速的连铸技术是水平连铸。尽管有许多优点，但仍存在两个主要问题，即重力效应和无法生产宽板坯。使用电磁搅拌可以成功地解决这两个问题。

电磁搅拌工艺的主要缺点是"白亮带"的出现。这是负偏析的结果，因为它在硫印或酸蚀样品上看起来很多。当使用单个强力搅拌器时，问题更加突出。缓解该缺陷的途径包括限制搅拌功率、组合搅拌和三维搅拌。

正如本节所述，连铸过程中电磁搅拌的应用为表 6-3 所示的所有问题提供了解决方案。据报道，人们应该对通过电磁搅拌获得的铸坯质量和性能的实质性改善充满信心。这种技术有助于将连铸转变为能够生产高质量钢材的工艺，并将在该行业中发挥越来越重要的作用。

7 特殊连铸工艺

7.1 水平连铸

连铸已经从完全垂直的连铸机发展到低头/多点矫直铸机，其主要优点是降低了连铸机高度和钢水静压力。立式铸机的钢水静压对支撑辊和支撑段施加了剧烈的压力。很多研发集中在通过完全水平浇注实现几乎为零的钢水静压。然而，这涉及从垂直进料结晶器到水平进料结晶器的研发，需要水平中间包/结晶器接头和特殊手段以减少结晶器摩擦，因为结晶器通过接头刚性地固定到中间包上。

结晶器/中间包接头由一块耐火材料制成，称为分离环。根据铸机制造商的不同，中间包、水口、分离环和结晶器的布置会略有不同。分离环由特殊耐火材料制成，例如氮化硼或氮化硅（Si_3N_4），它必须耐热冲击、耐腐蚀并且不被钢润湿。此外，该部件加工尺寸非常精确。图 7-1 所示的是水平铸机的横截面图，其中中间包和结晶器刚性固定。水平连铸工艺具有以下优点：

（1）非常低头的连铸机，可以安装在普通的建筑物中；

（2）铸机设计提供全面防护，防止大气污染，特别是对于小断面，能够以小断面铝镇静钢；

（3）铸坯不变形，适用于特种钢连铸，如工具钢和高合金钢。

这不是一个全新的工艺，第一个浇注实验企业可以追溯到 1966~1967 年（英国的 Davy-Loewy，美国的通用汽车公司），但它近年来已经取得大量经济效益。已经开始了研究和开发工作，力图将这一工艺推向工业应用。自 1975 年以来，大约有 30 家企业建成；其中大多数仍然是由各种铸机制造商运营的试验企业，但是一些铸机已达到工业生产水平，例如 NKK Fukuyama（1978）、Boschgotthardshiitte（1980）、Armco（1984）和 British Steel（1988）。根据拉坯机制，水平铸机可分为两种类型：

（1）结晶器中间包组件静止，并且采用间歇拉坯模式。图 7-2 显示（a）中间包结晶器布置，（b）典型的拉坯模式和（c）坯壳的形成。这是最常用的设计，Technica Guss 和 Nippon Kokan/Davy Loewy 可能提供了迄今为止的大部分铸机。

（2）结晶器中间包组件振动并采用连续拉坯。这种技术已在前苏联和克房伯采用。

还应该引证两个原始的发源：现在似乎已经废弃的 Watts 工艺和俄罗斯开发的一个结晶器同时供给两个水平相对的铸坯（VN Ⅱ Metmach）。

水平铸造的铸坯尺寸为：

（1）直径 3~12mm 的线材，12 流铸机的产能可达到 25000 万吨；

（2）直径达 330mm；

（3）小方坯和大方坯 50mm 见方到 250mm 见方、130mm×170mm。

到目前为止，分离环技术的限制不允许用于板坯或大型坯铸造。拉坯速度与传统铸机上的拉速相似，但随着密集研发，拉速可能会提高。结晶器在连铸期间不润滑，由具有高抗腐蚀性和抗热变形性的铜合金制成。结晶器出口处通常有石墨部分。

这些工艺遇到的主要问题是：

（1）铸坯直径达 330mm；

（2）分离环的寿命和成本一般将铸造时间限制在几个小时；

（3）表面质量必须没有与凝固标记相关的横向裂纹（也称为冷疤或连系痕），这可能需要进行表面处理；

（4）难以向结晶器中供应润滑剂以减少摩擦；

（5）尽管类似的有色金属铸造工艺相当成熟，但它不适用于带钢的大断面。

如图 7-2（c）所示，凝固坯壳的形成是钢液在结晶器中分离环处凝固产生的。结晶器浸入式水口、分离环形状、流钢砖形状、振动或脉冲频率等技术发展以及对工艺（过热度，与钢种相关的结晶器锥度）的改进提高了铸坯表面质量。为获得最佳表面质量，必须调整不同钢种的结晶器振动周期。

就中心偏析和疏松而言，内部质量与常规铸坯类似，但在大断面铸坯上，组织呈现出一些不对称性。二冷区电磁搅拌是改善内部质量的另一个途径，消除了不对称组织。

该工艺起初主要用于特殊钢，与铸锭工艺相比，产量优势很大，但设备吨位要求和资本成本适中。若要拓宽应用范围，分离环的开发是至关重要的，分离环的成本和使用寿命可能是主要限制因素。这限制了铸机的生产率和该过程的经济效益。虽然高质量的耐火材料正在迅速发展，但在该工艺被更广泛接受之前，耐火材料技术仍然需要相当大的进步。

7.2 异型坯连铸

第一台异型坯连铸机在 30 年前投入运行。当时，连铸工艺本身处于起步阶段，许多所谓的专家质疑该工艺是否首先具有成本效益，其次，工艺是否能够生产所需质量的铸坯。

在过去的 30 年中，连铸工艺发生了巨大变化，生产钢材的理念也是如此。根据英国钢铁研究协会初步的异型坯连铸试验，阿尔果马钢公司在康卡斯特公司的支持下，决定安装一台异型坯连铸机。双流异型坯连铸机的初始设计基于五种不同的异型坯，覆盖了所需的成品范围。该项目的成功使用近终形铸坯生产树立了榜样。

通过连续铸钢生产宽丁字钢梁所谓的"厚边轧制"的成功在 1970 年代早期被日本钢铁工业迅速认可。表 7-1 概括了日本异型坯连铸机的发展。

川崎钢铁从康卡斯特公司订购了第一台大型连铸机。该连铸机在美国设计并在日本制造，推动了近终形工艺的全面工业应用。

其他生产钢梁的日本公司也认可了成本效益，导致后续的铸机均基于康卡斯特的技术制造。

日本由于建筑业大幅发展，并位于一个严重的地震带，对钢梁的需求在 20 世纪 80 年代开始大幅增加。这种增长逐步升级，如今的需求量每年超过 800 万吨，大约是北美需求量的两倍。

由于对重建基础设施的需求极少，目前美国对钢梁的需求并不大。钢梁的供应多年来主要由大型钢铁联合企业生产，特别是伯利恒钢铁公司、美国钢铁公司和内陆钢铁公司。

小型企业在1970年代生产棒材方面取得了巨大成功。其中三家小型钢厂认识到他们也可以成为较小的H和I型钢梁的生产商，并且能够以远低于三家生产商的成本生产这些产品。这引起纽柯公司于1988年与日本大河钢公司成立合资企业，最初开发的技术现在可以快速付诸实施。

纽柯-大和钢公司位于阿肯色州Blytheville的结构钢生产设施由康卡斯特提供三流异型坯连铸机，可生产宽达27in的丁字钢梁。使用的轧机基本上是传统的轧机，但是，劳动力成本非常低，并且连铸机和轧机的效率使其相对于大型生产商具有巨大的成本优势。

1986年，康卡斯特改造了西北钢铁公司现有的六流铸机，生产小型轻型钢梁坯。

人们开发了适用于这些异型坯的管式结晶器。与此同时，查帕拉尔钢公司的五流铸机也被转换为使用管式异型坯结晶器来铸造小断面铸坯。这部分铸机将用于中型轧机，并生产小型轻质梁。

铸造工艺的成功发展主要取决于对传输现象基础知识的理解和应用。传热和应力对铸坯质量有非常重要的影响，如裂纹形成、铸坯组织和性能等。多年来，人们对连铸中的传热和应力进行了大量的研究，为实际生产提供了许多有用的信息。然而，这些模型仅限于简单几何形状的连续铸造，例如小方坯、大方坯和板坯。在异型坯的连续铸造中，主要由于其复杂的形状，可能发生各种缺陷。从其形状、温度和应力分布可以看出，异型坯连铸中的凝固模式与简单形状的铸造中的凝固模式不同。遗憾的是，文献中关于近终形铸坯模拟的信息很少，特别是在热-机械性能方面。

7.3 薄板坯连铸

在过去十年中，来自美国、欧洲和日本的钢铁生产商大力开发研究直接浇注工艺。人们的兴趣源于简化整个炼钢过程来降低生产和投资成本的意愿。这种简化是由于钢材的连铸接近或达到冷轧所需的厚度。在连铸和冷轧之间消除大多数工艺步骤简化流程，可节省能量并降低投资成本。直接铸造研究主要集中在三种传统工艺配置上：双带式、双辊式和单辊式布置。此外，相当多的研究正致力于新的创新，如喷射成形和电磁悬浮铸造工艺。改进双带式工艺是为了铸造厚度约为25~75mm的薄板坯，而对于双辊式和单辊式工艺，重点是连铸通常为1~6mm和20~500mm的薄带。薄板坯通常不够薄，不能直接引入冷轧设备，并且通常需要一些额外的热轧。

薄板坯连铸可以提供近终形状的铸坯，并且仅需要较少的轧制过程，因此与传统的板坯连铸相比更加经济。在薄带钢连铸工艺中，钢液位于在两个旋转辊和两个固定的耐火侧板之间。所获得的板的形状和尺寸由中间包中的辊直径和钢液深度决定。由于局部温度差异，靠辊滑动的固定侧板承受高应力。板材温度从钢铁的熔化温度（1500~1600℃）到轧辊接触和背面坝温度（200~400℃）不等。

7.3.1 存在的问题

（1）与传统的板坯连铸相比，薄板坯连铸的拉坯速度大于6m/min，结晶器的宽窄边比值大，因此结晶器中的钢液流动很难防止坯壳初始凝固不均匀、弯月面的剧烈振动以及铸造过程中的一些其他缺陷。钢液流动或钢液流动的几种数值模拟与其他现象相结合，在之前发表的研究中可以找到详细的资料。然而，这些研究中的大多数是基于具有矩形或圆形断面的简单几何系统，因此需要简便的流体动力学研究。Honeyands 和 Herbertson 利用水模型对薄板坯连铸结晶器进行了先驱性流体动力学研究。为了减少钢液供给的湍流及其对结晶器壁的冲击，薄板坯连铸结晶器设计为漏斗型。相应地，SEN（浸入式水口）专门设计带有出口部分。Nam 等人使用基于结构化拟合坐标网格的有限体积法在漏斗型薄板连铸结晶器中进行流体流动与传热和凝固相结合的数值分析，分析了结晶器中的传输现象。此后，Park 等人对平行板型薄板坯结晶器进行了数值分析，并且对 SEN 的优化设计给予了更多的关注，这对钢液流动模式、传热和凝固有相当大的影响。

（2）在分析该过程时，侧封板磨损不是唯一需要考虑的问题，还有局部温差引起的热冲击和高应力；侧封板与钢液之间的化学反应；并且在导辊上意外凝固的坯壳可能导致侧封板快速损坏。在轧辊接触区域中产生最大侧封板拉伸应力。在侧封板的底部可能发生裂纹，导致钢液渗入接触区域并导致侧封板的腐蚀。

这种腐蚀必须通过导辊对耐火侧封板的机械磨损来补偿，以通过保持导辊和侧板之间的紧密接触来避免这种渗透。根据实验知识，通过侧封板的恒定位移获得机械磨损。这种方法导致侧封板的大量消耗并限制了产能。

这种机械磨损必须从承载负载的侧封板控制。如果耐火材料磨损取决于侧封板和导辊之间的接触压力，则可以应用该方法。机械磨损过程的控制需要理解第三相形成机制和速度调节机制（VAM）。磨损控制还需要确定 SiAlON-BN 磨损的影响参数及其量化。

7.3.2 实际生产

目前，在汽车工业中使用铝是一种成熟的做法，使制造商能够提高车辆燃油经济性并减少二氧化碳排放。铝与竞争材料相比，优势在于其低密度、高强度和塑性、易于回收和高耐腐蚀性的组合。虽然铝压铸件的应用范围很广，但铝板的使用受到限制。在汽车工业中广泛使用铝板的主要障碍是其成本高，是钢板的 4~5 倍。通过双辊连续（TRC）铸造而不是通过传统的直接冷却（DC）铸造和热轧生产铝板提供了大幅降低板材成本的机会。

由于在 TRC 铸造中的高凝固速率，TRC 铸造材料的微观组织与 DC 铸造材料的微观组织明显不同。TRC 合金通常在固溶体中表现出高浓度的合金元素、细小的初级金属间化合物颗粒和细的铸态晶粒尺寸。所有这些特征都影响合金对下游加工中涉及的热机械处理的响应。因此，由 TRC 和 DC 铸造合金生产的板的微观组织可以显著不同。微观组织和晶体结构的差异对薄板材的机械性能和塑性有很大影响。

用于汽车的铝板应具有良好的塑性和足够的强度（除了其他性能外）。目前使用的 DC 铸造材料相当好地满足了这一要求。最近，人们越来越关注使用 TRC 作为生产用于汽车结构部件的低成本/高质量 Al-Mg（AA5xxx 系列）板的方法。TRC 铸造 Al-Mg 合金的研究工作集中在确定下游加工的参数上，以获得具有与 DC 铸造合金相当或更好性能的产品。因此，需要详细了解由两种铸造方法生产的板材的微观组织，包括它们的晶体结构。

7.4 带钢浇注工艺

双辊带钢浇注工艺在名为 Myosotis 的项目下开发，可以通过图 7-3 所示的双辊带钢浇注来说明。在两个镀镍铜辊之间引入钢液。与轧辊接触的两个耐火侧板容纳钢液。辊是水冷的，直径为 1.5m，长度为 1m。两个辊之间的间隙决定了钢板的厚度。在研究开始时，这个新工艺能够在一次浇注约 90t 钢液，生产长 3.5km、宽 0.865m、厚 2.8mm 的钢板。侧封板的接触区域由维苏威公司开发的 SiAlON-BN 复合材料组成。

双辊薄带连铸被认为是近终形连铸的新一代技术。所生产的薄带厚度约为 1~6mm。薄带连铸虽然具有能耗低、投资少、排放少的优点，但存在运行平稳、铸坯质量等问题。

在带钢浇注期间，熔池中的流场模式和液面波动影响带材质量。当液面波动超过 ±1.6~2mm 时，将产生纵向裂纹。相反，如果弯月面过于平静，则熔池的下部会增强湍流，表面层会凝固。此外，进料系统的配置、浇注速度和熔池深度也会影响流场。因此，控制熔池中的流场非常困难。目前，影响钢带质量的关键因素是辊缘、供料系统和辊特性。进料系统的配置对熔池中的流场具有决定性的重要影响。必须在熔池宽度范围上均匀地供应钢液，因为不对称的流场模式对凝固产生负面影响。流场模式也受到供料系统几何形状的影响，并且由两个辊表面形成的结晶器中的传热也受到流场模式的影响。

几年来，研究主要集中在如何节省能源、时间和成本以及如何获得具有新特性的新材料等问题上。关于扁平材料，双辊薄带铸造是一个可行的途径。这是一个复杂的过程，最终铸坯是 2~7mm 厚的热轧带钢。工艺的复杂性是在一个步骤中结合凝固和成形而导致的。必须在铸造的小部分时间内满足最终产品的所有要求。

IBF 的连铸机与蒂森克虏伯的连铸机一起运行，由感应炉、两个辊轴平行的双辊连铸机、夹送辊和卷取机组成。如果需要，可以将喷水冷却区添加到工艺中。通过浸入式水口在轧辊之间浇注钢液。陶瓷侧板密封正面。凝固发生在旋转的辊处，形成带坯壳的薄板。它们在最窄的辊缝之间或之上的接合点处合并，完全凝固的薄带离开辊缝。

7.4.1 双辊式浇注

尽管最早涉及双辊式浇注的专利是 20 世纪中期生产金属带的方法，但是该工艺足足花费了一百多年的时间用于商业开发。在 20 世纪 50 年代早期，亨特工程公司推出了一种用于生产铝带的双辊连铸机。在这些早期的铸机中，水冷的轧辊水平排列，并通过耐火进料器尖端从下方进料。最早的铸机的辊直径为 600mm，并且辊以约 600t 的液压力夹在一起。由这类铸机生产的薄板带通常为 6mm 厚、1500mm 宽。

铸坯离开辊后，在进入包括一对夹送辊、静态剪切机、四辊式张紧装置和最后一个卷取机的带材处理设备之前，使带材成卷。虽然这类铸机被广泛用于铝箔材料的生产，但是它们在其他应用中受到许多工艺和设备的限制。

自引入亨特立式连铸机以来，设备和铸造技术逐渐发展，但基础工艺技术几乎没有变化。目前正在使用具有较大辊（通常为 1000mm 直径和 2000mm 宽）的铸机，并且通常将辊垂直布置或略微倾斜布置，使得板坯从轧辊上水平地生产。这类铸机可以铸造一系列合金，但通过这种方法生产的绝大多数材料是用于铝箔坯料或翅环孔型坯等洁净（窄凝固范围）合金。尽管设备相对简单并且缺乏有效的工艺开发，但双辊连铸机铸坯适用于下游加工，许多操作人员生产厚度为 7μm 或以下的成品箔坯。

7.4.1.1 工艺过程

从设备的角度来看，双辊连铸机本质上是一种缓慢的双辊轧机，原料是熔融铝，最终产品是 6~10mm 厚的铝带。当钢液与水冷辊接触时开始凝固，并且由于轧辊咬合的尺寸逐渐减小，凝固金属被迫保持与辊接触。一旦凝固完成，材料在离开夹持辊之前经历适当的热加工。

根据操作参数，在双辊铸造期间可以产生显著的分离力，并且由于这些高负荷，双辊铸造板的特征在于良好的表面质量和、铸坯规格和轮廓公差。即使对于诸如 AA 1145 合金之类的相对高纯度的材料，0.5t/mm 的载荷也是常态。对于宽凝固范围的合金，例如 AA 5182，荷载接近 1t/mm 常见的。由于这些高负荷，观察到辊的显著变形。举例来说，当铸造 1500mm 宽的 AA 1145 合金时产生的载荷足以使辊偏转达 2mm。轧辊中的这种偏转的结果是，为了产生标称 6mm 的薄板带，轧辊边缘处的初始辊隙必须设定为约 4.5mm。辊弯的影响传递到每个辊的直径上通常为 0.75mm，而辊中心线处的辊隙通常约为 3mm，这会引起显著的操作问题。

7.4.1.2 设备发展

铸机设计改进的实例是在铸造期间设定和调节辊缝的方法对铸造操作工具有显著益处并且已经改善了铸坯的质量。在最早的铸机上，通过在轴承座之间插入垫片来设定辊缝，并用塞尺测量。设定间距是一个困难的过程，一旦连铸正在进行，就无法调整间距。只有通过调整铸机一侧或另一侧的垫片尖端，才能克服薄板带中的侧向厚度变化，但这可能会导致其他问题。在后来的铸机上，简单的机械垫片被可调节的斜楔系统所取代。利用该系统，至少原则上可以在连铸期间调节辊缝，但是该过程要求液压力基本上减小到零以改变轴承座。就其本质而言，此程序是"命中或未命中"事件，需要多次尝试来纠正任何缺陷。当载荷传感器与斜楔调整结合使用时，戴维在 80 年代中期取得了重大进展。

铸造薄板产生的分离力是带材厚度和主要操作参数（例如拉坯速度和尖端缩回）的函数。如果可以精确地测量分离力，则不仅可以控制带钢的侧向厚度，而且可以控制薄板带的轮廓。板材的规格和轮廓公差变得越来越苛刻，这使铸机制造商和操作员的态度发生了根本性变化。通常，铸造带材中的侧向尺寸变化在 1% 以内，并且带材轮廓变化在 0%~1% 之间。

施加的夹持力与铸带中产生的分离力相反，合力-轴承力使用位于轴承座之间的力传感器测量。为了生产平行带钢，铸机每侧的轴承座受力需要相同，并且为了产生所需的产品轮廓，总分离力需要在预定限度内。

随着分离力从零增加到稳定值，辊隙的形状和尺寸改变并且薄板带轮廓从负通过平行变为正。所需的轮廓在平行和正 1% 之间，并且为了保持这种轮廓，必须将分离力控制在严格定义的极限之内。

使用载荷传感器测量分离力的瞬时值，并且可以将其用作控制系统的测量变量。基于连续测量分离力的三个回路，铸机控制系统在许多工业化铸机上运行。尽管这种发展是早期系统的重大进步，但在铸造期间改变轧辊间隙仍然很困难，并且通常仅在所有其他途径都已行不通时才尝试。最终的解决方案是完全更换斜楔组件，并调节气缸以液压方式设置连铸辊的间隙。使用这种技术，只需通过改变液压缸的信号，就可以毫不费力地实现辊缝的变化。铸机控制策略基本上保持相同，唯一的变化是不测量轴承座之间载荷传感器的分离力，而是通过位于液压缸中的压力传感器来测量。通过位于轴承座或轧辊之间的传感器或通过位于液压缸中的传感器测量辊隙。

7.4.1.3 实验方案

自 1989 年初在牛津大学安装实验性双辊连铸机以来，作为 SERC/DTI 支持的教学计划的一部分，已经试验了 600 多个铸机。按照最初的设想，连铸机包括一对直径为 400mm，表面宽为 300mm 的内部水冷钢辊，使铸带方向水平，只驱动底辊，并在开浇前手动设置辊缝。其他的报告说明了设备和研发计划的详细情况。

该过程的数值模拟预测，随着带钢厚度的减小，生产率提高，并且初始铸造试验旨在建立连铸机的操作能力并检查建模结果的有效性。如前所述，轧制偏转和配合拱度使得宽铸机的开浇变得困难，并且这个问题在薄规格薄带上被放大了。由于铸辊的固有刚度，对于窄宽度实验连铸机来说不是一个严重的问题，尽管由于仅驱动底辊，向非驱动辊的运动传递太慢而不能在薄的情况下启动没有金属过早凝固的量规。

为了克服这个困难，对实验铸机进行了修改，以结合液压间隙控制、每个轧辊的独立驱动和杠杆致动的尖端工作台。通过这种布置，可以在标称 4mm 规格下启动铸机，并在铸造期间逐渐减小辊隙。通过相应地提高铸机速度，可以高达 15m/min 的速度生产薄至 1mm 的带材。结合杠杆致动的工作台以便在辊隙减小的同时缩回尖端，同时保持尖端和两个辊之间的有效密封。

薄带铸造的成功取决于许多过程变量的控制。对于每种合金和标准，已经发现成功的铸造只能在设定的极限-操作窗口之间实现。影响该窗口的因素尤其是拉坯速度、尖端缩进、金属静压头、熔体过热、晶粒细化、带材张力、轧辊润滑、轧辊冷却和轧辊表面状况。每种合金的操作窗口的大小取决于缺陷的形成，例如热线、中心偏析、粘连和热撕裂以及薄板带的机械和微观组织特性。合金的铸造能力可以根据该窗口的相对尺寸来测量。已经发现铸造能力随着厚度的减小而降低，因为该过程变得更容易受到过程变量微小变化的影响，并且对于许多合金而言，该过程在薄规格上的稳定性有限。结果，手动操作甚至简单的过程自动控制变得越来越困难，有必要开发一种能够解释铸造过程的状态并控制过程变量以保持带材规格、质量和生产率均处于最佳水平的控制系统。

已经使用实验连铸机研究了各种合金。虽然主要是铝基，但合金的范围很广，从超高纯度材料到宽凝固范围的 Al-Sn、Al-Cu 和复合合金。

通过这些研究发现，一般来说，宽凝固范围合金的操作窗口明显大于窄凝固范围合金的操作窗口，使得它们对于理解高速薄规格的铸造工艺特别有用。过程变量对 AA 5182 铸件影响的详细研究表明，铸造过程如何受到铸造条件选择的影响，以及如何通过数学模型预测的较薄规格铸造可以实现生产率的提高。由于这种理解，已经证明可以将窄和中等凝固范围的合金铸造成厚度小至 1mm，而拉速高达 15m/min。这些限制是由于连铸机的机械设计而不是工艺问题，并且当完成当前一轮铸机修改时，可以设想进一步提高生产率。

7.4.1.4 粘连

在双辊铸造期间，带钢黏结到一个或两个铸辊上是很常见的。

如果粘连微乎其微，则可以继续铸造，尽管由于在粘连期间改善的传热导致带钢厚度的增加，铸坯材料可能必须被废弃。在严重粘连的情况下，连铸通常必须终止，因为带材牢固地粘在辊上，不能被移除。从辊上去除"粘连"的带钢总是会导致轧辊表面的损坏，在极端情况下需要重新研磨轧辊。为了使这个问题最小化，标准做法是用润滑剂连续喷涂轧辊表面。已经尝试了各种各样的润滑材料，但最广泛使用的润滑剂是石墨乳在水中的悬浮液。通常通过气雾喷嘴喷涂润滑剂，这些喷嘴可以往复运动或横越铸机辊。即使将润滑剂喷射到铸辊上仍然会发生粘连，并且在这种情况下通常的做法是减小拉速直至粘连变得易于控制。

从在牛津的实验铸机上进行的工作可以看出，粘连现象是高速带钢铸造的主要障碍。此外，在高速和薄规格带钢连铸下观察到的粘连特性与在传统速度和规格下观察到的不同。

可归纳如下：

（1）4~6mm 厚的带钢，在高拉速下发生黏着：即"热"铸造条件。增加润滑剂的喷涂速率或降低拉坯速度以提供最大的拉速，不会粘连。如果不发生粘连，则可能产生热线，这是与双辊铸造相关的另一个特征缺陷。虽然没有关于哪种合金会粘连的硬性规则，但在传统的双辊铸造中，一般观察到粘连趋势随着合金含量的增加而增加。

（2）当带钢的厚度减小到 1~2mm 时，粘连更明显。凝固范围大于约 50℃ 的合金即使在高拉坯速度下也不会粘连：例如拉速为 15m/min 的情况。当拉速降低时，具有中间凝固范围的合金倾向于粘连，而当拉坯速度增加时，窄凝固范围的合金倾向于粘连。

（3）当发生粘连时，前滑消失，前滑是带钢速度和铸辊圆周速度之间的差值，这在正常铸造条件下已经发现，通常为 5%~10%。

（4）在粘连开始时辊温度增加，并且带钢温度相应降低，带钢厚度（和分离力）增加。

（5）润滑剂太多会导致表面质量差和其他铸造缺陷，而润滑剂太少会导致粘连。润滑剂的喷涂率是关键，并且必须接近于其消耗的速率。

为了克服粘连问题，特别是在高速和薄规格铸造时，提出了许多解决方案：

（1）机械方法。尽管现有的喷涂装置足以用于传统的铸造（低速和薄的窄凝固范围合金），但它们已经被证明在高速薄带铸造条件下使用有限。主要原因是涂层在整个带钢宽度上分布不均匀，这是应用方法导致的结果。

脱模剂通常以横向或往复式喷嘴雾化喷涂到铸辊上。上述两种系统都没有将均匀的涂层沉积在连铸辊上。对于两种系统，润滑剂以螺旋形式沉积，螺旋的螺距是辊的旋转速度和喷嘴横向速率的函数。因此，辊的一些区域保持未涂覆，而其他区域，特别是在横梁的末端附近，被涂覆双层涂层。涂层明显缺乏均匀性是有害的，因为分离层的厚度影响从带钢中吸收热量的速率。在未涂覆的区域中存在粘连的可能，而在润滑剂过量的区域中，热量发散的速率减小并且可能导致诸如热线等其他铸造缺陷。为了克服这个问题，已经开发了一种改进的施加润滑剂的方法。

常见的结果是，当发生粘连时，前滑从大约 10% 减小到零。这意味着在正常稳态"非黏性"铸造期间，带钢速度比辊速度大 10%，一旦发生粘连，则带钢速度被迫和辊速相同。已经提出如果带钢可以总是具有相对于铸辊的擦洗运动，则不会发生粘连。达到这种目的可以通过改变辊的布置来实现，使得铸辊相对交叉而不是保持平行。通过交叉辊，产生与铸造方向成直角的滑移分量，并且仅需要非常少量的交叉来产生通常与"非黏性"铸造相关的相对运动量。轧辊交叉是轧机的成熟技术，用于控制带钢轮廓，虽然增加了铸机机械设计的复杂性，但潜在的优势使它值得应用。

（2）化学方法。对于高速薄板带连铸，已证明在凝固后期，薄板带表面上存在残余钢液有利于避免由于自润滑效应而导致的粘连。这在宽凝固范围合金中发生，但在使用双辊技术生产的大多数合金中不会发生。此外，这些合金的成分必须保持在严格限定的范围内，以符合国际公认的标准。因此，问题可归纳如下："合金的有效凝固范围必须大幅提高，而不会使合金成分超出规格"。当人们考虑到这一点时，尽管精确的数值将取决于所讨论的合金，但需要对成分进行重大改变才能使凝固范围发生适度的变化，除非可以找到一些"神奇"成分，否则这项任务似乎是不可能完成的。

对铝合金相图的检验证实存在这样的合金元素并且候选材料的实验已经证明，通过微量添加特定的合金元素可以显著改善粘连。

（3）工艺解决方案。对于 4~6mm 厚的带钢，克服粘连的传统方法是逐渐降低拉坯速度，直到粘连不再显著。减慢拉速（或者增加尖端缩回），分离力增加，这具有增加前滑程度的效果。带钢和辊之间的相对运动的增加阻止了粘连。

仅用一个辊驱动铸机，可以确定前滑在防止粘连中的作用。当仅驱动一个辊时，另一个辊由铸坯驱动。这意味着带钢以与从动辊（零前滑）基本相同的速度行进，而非驱动辊的行进速度比带钢速度慢，即带钢相对于非驱动辊滑动。在这些条件下操作，在非驱动辊上不会发生粘连，并且在铸造期间从动辊的外观保持不变，非驱动辊形成有光泽的表面涂层。从这些结果可以看出，轧辊的相对速度非常重要。

尽管双辊铸造是一种成熟的技术，但该方法在铸坯质量和生产率方面都存在局限性，并且仅用于有限的铸坯。对设备设计的改进改善了工艺的操作和控制，并使铸带的规格和轮廓公差得到显著改善。通过为期四年的实验计划，在此期间，使用各种材料进行了 600 多次试验，对该过程的理解程度得到了扩展，从而大大提高了生产力和能力。到目前为止，铸造合金被认为是"不可铸造的"。正在进行进一步的开发，其中包括在商业上证明高速带钢铸造的概念，以达到联合开发协议的要求。

7.4.2 高速钢薄板带连铸

直接由钢液生产带钢的带钢连铸工艺作为钢铁工业的新技术引起了人们的极大兴趣，原因在于带钢连铸工艺的发展不仅可以实现工艺的缩短和能源消耗的降低，也是难成型材料的近终形连铸手段。薄板带连铸的各种方法中最有希望的方法之一是双辊法。在过去的十年里，已经对这种方法进行了大量的实验和分析研究。此类工作主要集中在不锈钢和硅钢的带钢连铸上。目前，至少在五个国家进行了不锈钢带钢连铸的先导试验。然而，在高速钢的薄板带铸造方面仅做了很少的工作。

高速钢是非常重要的工具钢，用途包括制造钢锯片。用于钢锯片的薄板厚度通常小于 2mm。当使用常规技术生产钢板时，为了减小铸坯的厚度和破坏铸态碳化物，多次轧制和多次退火是不可避免的，因此生产需要很长时间。在这项研究中，双辊法用于生产带钢，以大大缩短生产流程，带钢的冶金质量、微观组织、重点是碳化物以及钢锯片的切割性能，正在研究中。

参 考 文 献

[1] 史宸兴. 连铸历史回顾与未来 [J]. 连铸, 2016, 41 (4)：1~11.
[2] World Steel Association Economics Committee. Steel Statistical Yearbook [J]. Brussels：World Steel Association, 2018.
[3] Christoph S. Trends in Continuous Casting of Steel—Yesterday, Today and Tomorrow [C] //Plenary Session 7th ECCC, 2011.
[4] Brian G Thomas. In Modeling for Casting and Solidification Processing [M]. Marcel. Dekker, New York, 2001：499~540.
[5] Irving W R. Continuous Casting of Steel [M]. London SW1Y 5DB, The Institute of Materials 1 Carlton House Terrace.
[6] 王令福. 炼钢设备及车间设计 [M]. 北京：冶金工业出版社, 2007.
[7] 朱立光, 李琨, 李曜光, 等. 板坯连铸结晶器铜板传热行为 [J]. 铸造技术, 2016 (4)：706~709.
[8] 文光华, 杨昌霖, 唐萍. 连铸结晶器内渣膜形成及传热的研究现状 [J]. 工程科学学报, 2019 (1)：12~21.
[9] 王德兴. 连铸二冷区换热系数测试及配水优化研究 [D]. 沈阳：东北大学, 2017.
[10] 蔡开科. 浇注与凝固 [M]. 北京：冶金工业出版社, 1987.
[11] 赵茂国. IF 钢钢液结晶器内初始凝固行为研究 [D]. 唐山：华北理工大学, 2018.
[12] 雷少武, 王立波, 李东华, 等. Q235B 钢矩形坯在结晶器中传热凝固行为的研究 [J]. 炼钢, 2019 (4)：43~52.
[13] 崔立新, 张家泉, 陈素琼, 等. 连铸板坯在结晶器内凝固行为的研究 [J]. 炼钢, 2003 (3)：22~25.
[14] 孙瑞, 熊玮, 王杰, 等. 基于多场耦合模型的连铸结晶器内钢液凝固行为分析 [J]. 铸造技术, 2018 (2)：276~280.
[15] 干勇. 现代连续铸钢手册 [M]. 北京：冶金工业出版社, 2010.
[16] José Renê de Sousa Rocha, et al. Modeling and Computational Simulation of Fluid Flow, Heat Transfer and Inclusions Trajectories in a Tundish of a Steel Continuous Casting Machine [J]. Journal of Materials Research and Technology, 2019：4209~4220.
[17] 吴铿. 冶金传输原理 [M]. 北京：冶金工业出版社, 2014.
[18] Jörg Peter. Designing and Modeling of A New Continuous Steelmaking Process [D]. University of Missouri-Rolla. 2006.
[19] 朱立光, 等. 现代连铸工艺与实践 [M]. 石家庄：河北科学技术出版社, 2000.
[20] 曹磊, 王国承, 赵洋, 等. 钢液铝脱氧早期形成的 Al_2O_3 夹杂 [J]. 辽宁科技大学学报, 2018 (2)：88~93.
[21] 王林珠. 铝脱氧钢中非金属夹杂物细微弥散化的基础研究 [D]. 北京：北京科技大学, 2017.
[22] 张炯. 钢液中非金属夹杂物碰撞聚集行为的实验研究 [D]. 马鞍山：安徽工业大学, 2016.
[23] 孙军. 板坯连铸机拉速计算和控制 [J]. 冶金自动化, 2018 (2)：49~53.
[24] 陈瑜. 高碱度低碳钢保护渣流动性研究 [J]. 河北冶金, 2019 (6)：12~16.
[25] Jian Chen, Feng He, Yongli Xiao, et al. Effect of Al/Si Ratio on the Crystallization Properties and Structure of Mold Flux [J]. Construction and Building Materials, 2019, 4 (261)：19~28.
[26] 蔡开科. 连铸坯质量控制 [M]. 北京；冶金工业出版社, 2010.
[27] 赵茂国, 肖鹏程, 朱立光, 等. 铸造过程中 DP590 钢大型夹杂物研究 [J]. 铸造技术, 2018 (2)：281~283.
[28] L Zhang, B G Thomas. Inclusions in Continuous Casting of Steel [C] //XXIV National Steelmaking Symposium, November, 2003.
[29] 翟俊, 刘浏. EAF+AOD+LF 流程冶炼 310S 耐热钢夹杂物控制 [J]. 钢铁, 2017 (5)：31~35.
[30] 高新军, 郭永谦, 徐刚. 板坯角部横裂纹综述 [J]. 河南冶金, 2019 (3)：19~21.

[31] 张宏亮, 冯光宏, 崔怀周. 大规格 Q420B 热轧角钢轧制中间道次开裂缺陷分析 [J]. 钢铁, 2019 (9): 73~78.

[32] Minghe Ju, Jianchun Li, Xiaofeng Li, et al. Fracture Surface Morphology of Brittle Geomaterials Influenced by Loading Rate and Grain Size [J]. International Journal of Impact Engineering, 2019, 133: 1~41.

[33] 王胜利, 汪洪峰. 连铸板坯内部裂纹的形成机制及控制实践 [J]. 连铸, 2019 (2): 53~57.

[34] 谢长川, 李富帅. 扁坯窄面鼓肚变形的分析和避免措施 [J]. 特殊钢, 2018 (5): 23~27.

[35] 赵晓宏, 孙莹杰, 曹亮亮. 热轧圆棒材内部孔洞缺陷原因分析 [J]. 河南冶金, 2019 (1): 14~16.

[36] 董继亮, 汪云辉, 侯明山, 等. 中板坯连铸大梁钢 610L 的质量控制研究 [J]. 山西冶金, 2017, (2): 14~16.

[37] 佐祥均, 袁钢锦, 阎建武. 连铸过程中方坯脱方的原因与预防措施的研究 [J]. 连铸, 2015 (5): 29~33.

[38] 孙康. φ600mm 特大圆坯的凝固传热过程及鼓肚变形的研究 [D]. 秦皇岛: 燕山大学, 2012.

[39] 孟阳. 板坯黏结漏钢原因与预防措施 [J]. 天津冶金, 2015 (1): 15~17.

[40] 胡新宇. 小方坯漏钢浅析 [J]. 中国金属通报, 2018 (1): 163~164.

[41] 王玉昌, 张家泉. 高速重轨钢洁净度与均质性控制关键技术 [J]. 中国冶金, 2015 (4): 7~11.

[42] 许长军, 胡小东, 胡林, 等. 中间包导流挡板设计与冶金效果 [J]. 炼钢, 2013 (1): 69~73.

[43] 钟云涛, 潘汉玉, Emmanuel A, 等. 中间包等离子加热和电磁搅拌复合技术的开发与使用 [J]. 连铸, 2016 (6): 37~41.

[44] 杨滨. 通道式感应加热中间包内夹杂物碰撞长大和去除行为 [D]. 沈阳: 东北大学, 2016.

[45] 毛斌, 陶金明, 孙丽娟. 中间包冶金新技术——离心流动中间包 [J]. 连铸, 2008 (2): 8~11.

[46] 陈刚, 周海龙, 喻林. 浸入式水口结构对板坯结晶器液面波动的影响 [J]. 炼钢, 2018 (5): 50~56.

[47] 刘毅. 结晶器内大型夹杂物卷入机理及控制基础研究 [D]. 唐山: 华北理工大学, 2017.

[48] 李鹏飞, 葛建华, 王明林. 连铸坯热送热装在节能减排中的应用 [J]. 铸造技术, 2018 (8): 1768~1771.

[49] 熊良友, 陈小龙, 庞通. 柳钢连铸坯热送热装实践 [J]. 柳钢科技, 2018 (1): 50~52.

[50] 卫广运, 张瑞忠, 郭子强, 等. 连铸坯压下技术发展与应用 [J]. 河北冶金, 2019 (4): 25~29.

[51] Haibo S, Liejun L, Jinghui W, et al. Coordinating optimisation of F-EMS and soft reduction during bloom continuous casting process for special steel [J]. Ironmaking & Steelmaking, 2017, 45 (8): 708~713.

[52] 吴存有, 周月明, 侯晓光. 电磁搅拌技术的发展 [J]. 世界钢铁, 2010 (2): 36~41.

[53] Zhang L, Hou Y, Guo X, et al. Effect of Electromagnetic Stirring on the Microstructure and Properties of Fe-Cr-Co Steel [J]. Materials, 2018, 11 (8): 1~11.

[54] 王璞, 李少翔, 陈列, 等. 电磁搅拌对大圆坯结晶器冶金行为影响的探讨 [J]. 钢铁, 2019 (8): 82~89.

[55] 郑智聪. 铸铁型材水平连铸拉坯工艺参数控制规律与自适应整定研究 [D]. 西安: 西安理工大学, 2019.

[56] 徐春杰, 徐锦锋, 赵振. 铸铁水平连续铸造技术及铸铁型材的应用 [J]. 铸造技术, 2017 (11): 2559~2564.

[57] Nikolai Z, et al. Wear of Roll Surface in Twin-roll Casting of 4.5% Si Steel Strip [J]. ISIJ International, 2000, 40 (6): 589~596.

[58] 张兴中. 我国连续铸钢技术的发展状况和趋势 [J]. 钢铁研究学报, 2004, 16 (6): 1~6.

[59] Nikolai Z. Comparison of Continuous Strip Casting with Conventional Technology [J]. ISIJ International, 2003, 43 (8): 1115~1127.

[60] 杨军, 董洁, 张从容, 等. 铸坯成型理论 [M]. 北京: 冶金工业出版社, 2015.

[61] 毛新平. 薄板坯连铸连轧微合金化技术 [M]. 北京: 冶金工业出版社, 2008.

[62] 张剑君, 毛新平, 王春峰. 薄板坯连铸连轧炼钢高效生产技术进步与展望 [J]. 钢铁, 2019 (5): 1~8.